THE BADGE OF A FIREFIGHTER IS THE MALTESE CROSS.

Dating back to the time of the Crusades, it is a symbol of charity, loyalty, chivalry, gallantry, generosity to firend and foe, protection of the weak, and dexterity in service.

Today, firefighters everywhere are proud to wear the time-honored Maltese Cross as part of their uniform.

The Maltese Cross and the Fire Service, American Fire Journal, v. 50 (4), p. 14, April 1998

FIREFIGHTER FATALITIES IN THE UNITED STATES IN 2004

U.S. Department of Homeland Security

Federal Emergency Management Agency

U.S. Fire Administration

August 2005

In memory of all firefighters
who answered their last call in 2004

To their families and friends

To their service and sacrifice

U.S. Fire Administration
Mission Statement

As an entity of the Federal Emergency Management Agency, the mission of the U.S. Fire Administration is to reduce life and economic losses due to fire and related emergencies, through leadership, advocacy, coordination, and support. We serve the Nation independently, in coordination with other Federal agencies, and in partnership with fire protection and emergency service communities. With a commitment to excellence, we provide public education, training, technology, and data initiatives.

On March 1, 2003, FEMA became part of the U.S. Department of Homeland Security. FEMA's continuing mission within the new department is to lead the effort to prepare the Nation for all hazards and effectively manage Federal response and recovery efforts following any national incident. FEMA also initiates proactive mitigation activities, trains first responders, and manages Citizen Corps, the National Flood Insurance Program, and the U.S. Fire Administration.

ACKNOWLEDGMENTS

This study of firefighter fatalities would not have been possible without the cooperation and assistance of many members of the fire service across the United States. Members of individual fire departments, chief fire officers, United States Forest Service personnel, the Department of Justice, National Fire Protection Association (NFPA), the National Fallen Firefighters Foundation, and many others contributed important information for this report.

C2 Technologies, Inc., of Vienna, Virginia, conducted this analysis for the U.S. Fire Administration (USFA) under contract EME-2003-CO-0282.

The ultimate objective of this effort is to reduce the number of firefighter deaths through an increased awareness and understanding of their causes and how they can be prevented. Firefighting, rescue, and other types of emergency operations are essential activities in an inherently dangerous profession, and unfortunate tragedies do occur. This is the risk all firefighters accept every time they respond to an emergency incident. However, the risk can be greatly reduced through efforts to improve training, emergency scene operations, and firefighter health and safety.

PHOTOGRAPHIC ACKNOWLEDGMENTS

The USFA would like to extend its thanks to individuals and organizations that provided photographs for this report. All uncredited photographs were taken at the National Fallen Firefighters Memorial weekend in Emmitsburg, MD, October 2004, by Patti Odbert, Scientific and Commercial Systems Corporation.

TABLE OF CONTENTS

continued on next page

BACKGROUND

For 28 years, the U.S. Fire Administration (USFA) has tracked the number of firefighter fatalities and conducted an annual analysis. Through the collection of information on the causes of firefighter deaths, the USFA is able to focus on specific problems and direct efforts toward finding solutions to reduce the number of firefighter fatalities in the future. This information is also used to measure the effectiveness of current programs directed toward firefighter health and safety.

The results of this effort have been used to direct specific efforts targeted at reducing firefighter fatalities. The USFA publications *Safe Operations of Fire Department Tankers* (FA-248) and *Emergency Vehicle Safety Initiative* (FA-272) were both inspired by the high numbers of vehicle-related deaths highlighted in previous editions of this report. Both publications are available free of charge to the fire service from the USFA in paper and electronic versions at *www.usfa.fema.gov/applications/publications/*

One of the USFA's main program goals is a 25-percent reduction in firefighter fatalities in 5 years and a 50-percent reduction within 10 years. The emphasis placed on these goals by the USFA is underscored by the fact that these goals represent one of the five major objectives that guide the actions of the USFA.

In addition to the analysis, the USFA provides a list of firefighter fatalities to the National Fallen Firefighters Foundation. If Memorial criteria are met, the fallen firefighter's next of kin, as well as members of the individual fire department, are invited to the annual Fallen Firefighters Memorial Service. The service is held at the National Emergency Training Center in Emmitsburg, Maryland, during Fire Prevention Week. Additional information regarding the Memorial Service can be found at *www.firehero.org* or by calling the National Fallen Firefighters Foundation at (301) 447-1365.

Other resources and information regarding firefighter fatalities, including current fatality notices, the National Fallen Firefighters Memorial database, and links to the Public Safety Officer Benefit (PSOB) program can be found at www.usfa.fema.gov/fatalities/.

INTRODUCTION

This report continues a series of annual studies by the USFA of onduty firefighter fatalities in the United States.

The specific objective of this study is to identify all onduty firefighter fatalities that occurred in the United States and its protectorates in 2004 and to analyze the circumstances surrounding each occurrence. The study is intended to help identify approaches that could reduce the number of firefighter deaths in future years.

In addition to the 2004 overall findings, this study includes two special topics related to healthful eating and operational changes that can have an immediate effect on firefighter safety.

WHO IS A FIREFIGHTER?

For the purpose of this study, the term firefighter covers all members of organized fire departments in all States, the District of Columbia, and the Territories of Puerto Rico, the Virgin Islands, American Samoa, the Commonwealth of the Northern Mariana Islands, and Guam. It includes career and volunteer firefighters; full-time public safety officers acting as firefighters; State, Territory, and Federal government fire service personnel, including wildland firefighters; and privately employed firefighters, including employees of contract fire departments and trained members of industrial fire brigades, whether full- or part-time. It also includes contract personnel working as firefighters or assigned to work in direct support of fire service organizations.

Under this definition, the study includes not only local and municipal firefighters but also seasonal and full-time employees of the United States Forest Service, the Bureau of Land Management, the Bureau of Indian Affairs, the Bureau of Fish and Wildlife, the National Park Service, and State wildland agencies. The definition also includes prison inmates serving on firefighting crews; firefighters employed by other governmental agencies, such as the United States Department of Energy; military personnel performing assigned fire suppression activities; and civilian firefighters working at military installations.

WHAT CONSTITUTES AN ONDUTY FATALITY?

Onduty fatalities include any injury or illness sustained while onduty that proves fatal. The term *onduty* refers to being involved in operations at the scene of an emergency, whether it is a fire or non-fire incident; responding to or returning from an incident; performing other officially assigned duties such as training, maintenance, public education, inspection, investigations, court testimony, and fundraising; and being on-call, under orders, or on standby duty, except at the individual's home or place of business. An individual who experiences a heart attack or other fatal injury at home as he or she prepares to respond to an emergency is considered onduty when the response begins. A firefighter who becomes ill while performing fire department duties and suffers a heart attack shortly after arriving home or at another location may be considered onduty since the inception of the heart attack occurred while the firefighter was onduty.

On December 15, 2003, the President of the United States signed into law the Hometown Heroes Survivors Benefit Act of 2003. After being signed by the President, the Act became Public Law 108-182. The law presumes that a heart attack or stroke are in the line of duty if the firefighter was engaged in nonroutine stressful or strenuous physical activity while onduty and the firefighter becomes ill while onduty or within 24 hours after engaging in such activity. The full text of the law is available in Appendix C.

The inclusion criteria for this study will be affected by this change in the law. Previous to December 15, 2003, firefighters who became ill as the result of a heart attack or stroke after going offduty needed to register some complaint of not feeling well while still onduty in order to be included in this study. For firefighter fatalities after December 15, 2003, firefighters will be included in this study if they become ill as the result of a heart attack or stroke within 24 hours of a training activity or emergency response. Firefighters who become ill after going offduty where the activities while onduty were limited to nonstressful tasks that did not involve physical exertion such as clerical, administrative, or nonmanual in nature, will not be included in this study.

A fatality may be caused directly by an accidental or intentional injury in either emergency or nonemergency circumstances, or it may be attributed to an occupationally related fatal illness. A common example of a fatal illness incurred onduty is a heart attack. Fatalities attributed to occupational illnesses also would include a communicable disease contracted while onduty that proved fatal, when the disease could be attributed to a documented occupational exposure.

Injuries and illnesses are included even when death is considerably delayed after the original incident. When the incident and the death occur in different years, the analysis counts the fatality as having occurred in the year in which the incident took place.

For example, a California firefighter died in 2004 as the result of a heart attack suffered in July of 2003. In addition to the California firefighter, the USFA was notified of the death of an Ohio firefighter in 2003 who was not known or included in the firefighter fatality report for that year.

Information about these two deaths is included in Appendix A of this report, but they are not addressed in the body of the report unless the death affects retrospective statistical comparisons.

There is no established mechanism for identifying fatalities that result from illnesses such as cancer that develop over long periods of time, which may be related to occupational exposure to hazardous materials or products of combustion. It has proven to be very difficult over the years to provide a complete evaluation of an occupational illness as a causal factor in firefighter deaths due to the following limitations: insufficient tracking of firefighters' exposure to toxic hazards, the often delayed long-term effects of such toxic hazard exposures, and the exposures firefighters may receive while offduty.

SOURCES OF INITIAL NOTIFICATION

As an integral part of its ongoing program to collect and analyze fire data, USFA solicits information on firefighter fatalities directly from the fire service and from a wide range of other sources. These sources include the Public Safety Officers' Benefit (PSOB) program administered by the Department of Justice, the National Institute for Occupational Safety and Health (NIOSH), the Occupational Safety and Health Administration

(OSHA), the United States military, the National Interagency Fire Center, and other Federal agencies.

The USFA receives notification of some deaths directly from fire departments, as well as from such fire service organizations as the International Association of Fire Chiefs (IAFC), the International Association of Firefighters (IAFF), NFPA, the National Volunteer Fire Council (NVFC), State fire marshals, State training organizations, other State and local organizations, fire service Internet sites, news services, and fire service publications. The USFA also tracks fatal fire incidents as part of its Major Fires Investigation Program and performs an ongoing analysis of data from the National Fire Incident Reporting System (NFIRS).

PROCEDURE FOR INCLUDING A FATALITY IN THE STUDY

In most cases, after notification of a fatal incident, the USFA makes initial telephone contact with local authorities to verify the incident, its location, jurisdiction, and the fire department or agency involved. Further information about the deceased firefighter and the incident may be obtained over the phone from the chief of the fire department or his or her designee or by other data collection forms.

Information that is requested routinely includes NFIRS-1 (incident) and NFIRS-3 (fire service casualty) reports, the fire department's own incident reports and internal investigation reports, copies of death certificates or autopsy results, special investigative reports, police reports, photographs and diagrams, and newspaper or media accounts of the incident. Information on the incident may also be gathered from NFPA or NIOSH reports on an incident.

After obtaining this information, a determination is made as to whether the death qualifies as an onduty firefighter fatality according to the previously described criteria. With the exception of firefighter deaths after December 15, 2003, the same criteria were used for this study as in previous annual studies. Additional information may be requested, either by followup with the fire department directly, from State vital records offices, or from other agencies. The USFA makes the determination as to whether a fatality qualifies as an onduty death for inclusion in this statistical analysis. The National Fallen Firefighters Foundation makes the final determination as to whether a fatality qualifies as a line-of-duty death for inclusion in the Fallen Firefighters Memorial Service.

Overleaf: The apparatus carrying the body of Firefighter Jaime Leah Foster proceeds through a ladder arch at her funeral.
PHOTO BY RICK MCCLURE, LOS ANGELES CITY FIRE DEPARTMENT

2004 FINDINGS

One hundred and seventeen (117) firefighters died while onduty in 2004. This level of fatalities continues a disturbing upward trend in the number of firefighter fatalities. Even if the horrible toll of September 11, 2001, is set aside momentarily, more firefighters have died each year onduty in the past several years than would be expected after the lower loss years in the 1990's.

In 2004, this total was affected by a change in the inclusion criteria for this report. In 2004, there were seven firefighter fatalities that would not have been considered for inclusion in this report prior to the enactment of the Hometown Heroes Survivors Benefit Act of 2003. This law generated a change in report criteria to include firefighters who die within 24 hours of stressful onduty activity.

With the deaths of 117 firefighters in 2004, this is the eighth time in the past 10 years, and the eleventh time within the past 15 years, when the total number of firefighter fatalities has reached or exceeded 100. The lowest years on record are 1992 with 77 fatalities and 1993 with 81 fatalities (Figure 1).

The 117 deaths in 2004 comprise almost 112 percent of the 10-year floating average and 108 percent of the 5-year floating average. The 117 deaths resulted from a total of 114 incidents. There were three firefighter fatality incidents where two firefighters were killed in 2004.

In 2001, 344 firefighters were killed as a result of the attacks on the World Trade Center (WTC) in New York City on September 11th of that

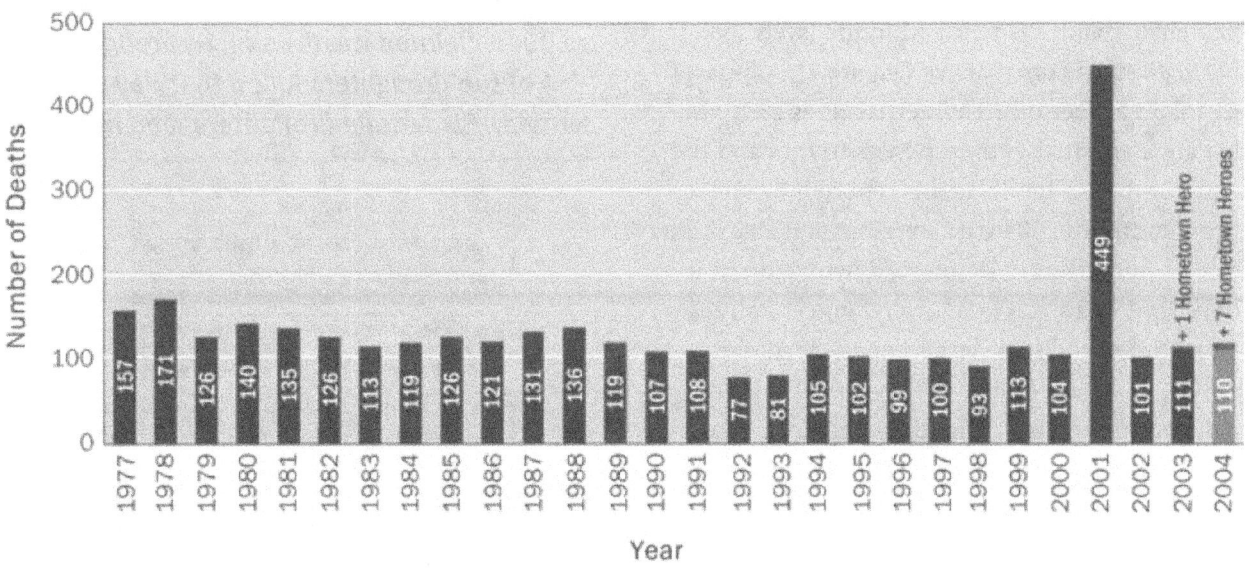

Figure 1. Onduty Firefighter Fatalities (1977-2004)

7

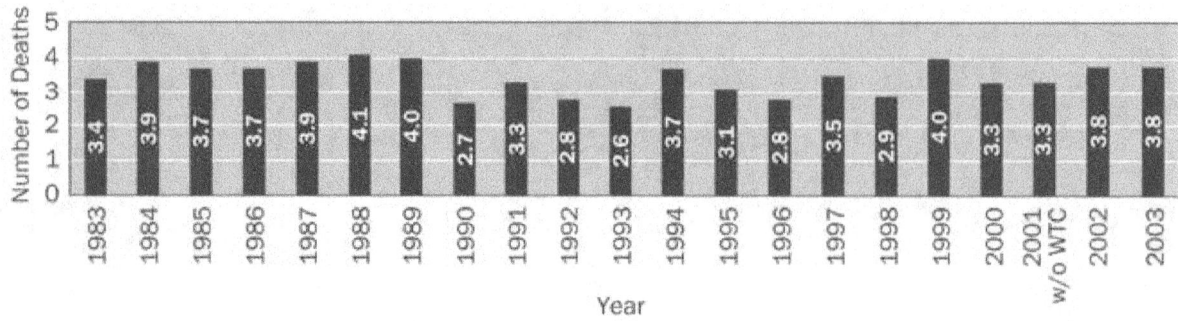

Figure 2. Firefighter Fatalities per 100,000 Fires

year. When conducting multiyear comparisons of firefighter fatalities in this report, it may be necessary to set these deaths apart for illustrative purposes. This action is by no means a minimization of the supreme sacrifice made by these firefighters.

The first onduty firefighter to die in 2004 was Lieutenant Leslie W. Gant, Jr. of New Jersey. He suffered a stroke after responding to a vehicle crash on a local expressway.

While the total number of firefighter fatalities has shown a downward trend over the past 20 years, the number of firefighter deaths per fire incident actually has returned to the levels seen in the 1980's. The chart above (Figure 2) compares the total number of firefighter fatalities each year that are associated with responses to fires and the total number of fire incidents reported by NFPA through 2003 (2004 data were not available at the time of this report). Despite a downward dip in the early 1990's, the level of firefighter fatalities is back up to the same levels experienced in the 1980's. If the firefighter deaths at the WTC are included in the 2001 data, the number rises to 23.1 firefighter fatalities per 100,000 fires.

Career and Volunteer Deaths

Firefighter fatalities in 2004 include 81 volunteer firefighters and 36 career firefighters (Figure 3). Among the volunteer firefighter fatalities, 72 were from local or municipal volunteer fire departments, and 9 were seasonal or contract members of wildland fire agencies. All of the career firefighters who died were members of local or municipal fire departments. Six of the firefighters who died in 2004 were female, and 111 were male.

The six female firefighter deaths in 2004 are exceeded only by the seven female firefighter deaths that occurred in 1994. The Storm King Mountain Fire on July 6, 1994, claimed the lives of 14 firefighters; 4 of the firefighters killed that day were female. Six female firefighters died in 1995.

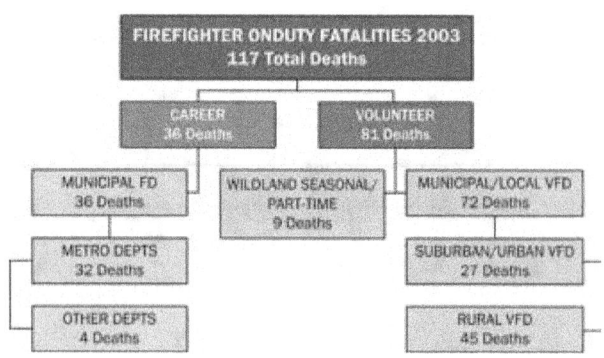

Figure 3. Career and Volunteer Firefighter Deaths

Multiple-Firefighter Fatality Incidents

The 117 deaths resulted from 114 incidents. There were three multiple-firefighter fatality incidents resulting in the deaths of six firefighters.

Table 1. Multiple-Firefighter Fatality Incidents

Year	Number of Incidents	Number of Fatalities
2004	3	6
2003	7	20
2002	9	25
2001	8	362
2001 w/o WTC	7	18
2000	5	10
1999	6	22
1998	10	22
1997	8	17
1996	3	8
1995	7	18
1994	6	26

Six firefighters died in three incidents that claimed the lives of two firefighters each. Two Pittsburgh firefighters were killed in the collapse of a burning church; two Wood River, Nebraska, firefighters

The three multiple-firefighter fatality incidents in 2004 resulted in six firefighter deaths. This is the lowest total number of multiple-firefighter fatality incidents and the lowest total number of firefighter deaths from these incidents in at least 10 years.

were killed in the collapse of a burning single-family residence; and two Philadelphia firefighters were killed when they became trapped in the basement of a burning home (the Philadelphia Fire Department also suffered the loss of a firefighter in a January structure fire).

Wildland Firefighting Deaths

Twenty-one firefighters died in 2004 while engaged in activities related to brush, grass, or wildland firefighting. This total includes part-time and seasonal wildland firefighters and municipal or volunteer firefighters engaged in fighting a wildland fire. This total is down sharply from the 29 firefighters who died engaged in similar activities in 2003.

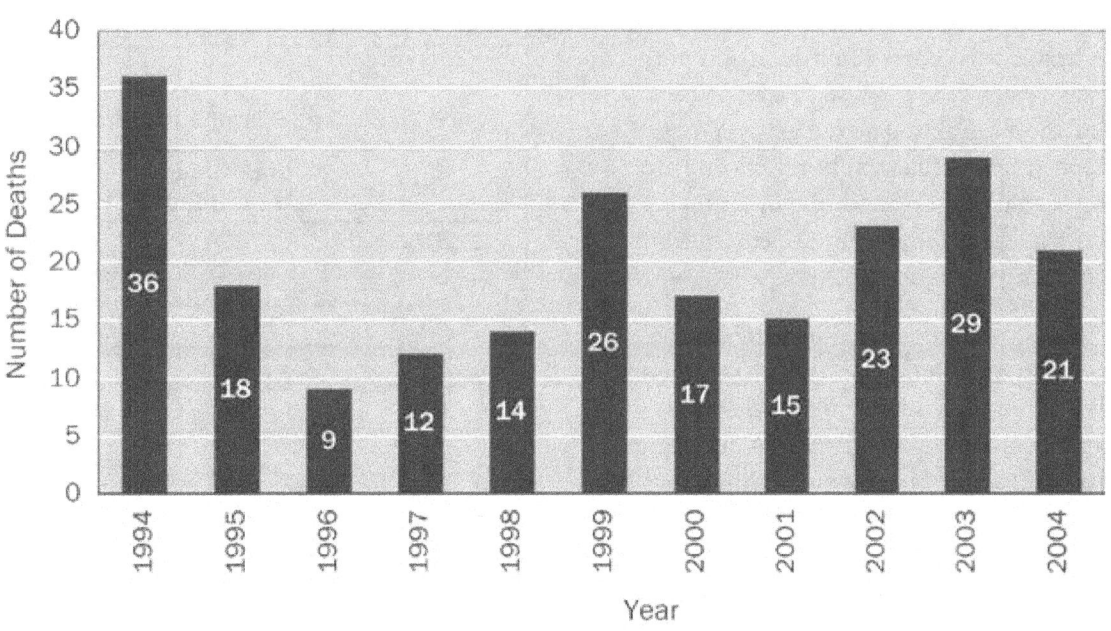

Figure 4. Firefighter Fatalities Related to Wildland Firefighting (1994-2004)

While the total number of firefighters killed in association with wildland fires was lower in 2004 than in previous years (Figure 4), there were no multiple-firefighter fatality incidents in 2004 that were related to wildland firefighting activities. The number of fatal incidents related to wildland firefighting has been exceeded only once in the past decade. While the total number of firefighters killed in association with wildland firefighting in 2004 is lower than previous years, the high number of fatal wildland incidents in the past two years is of concern.

Five firefighters died in vehicle crashes in association with wildland fires: two Mississippi firefighters were killed in separate incidents while responding to brush fires, one in a tanker and one in a brush/rescue truck; a Kentucky firefighter was killed when the engine that she was driving to a wildland fire crashed into a tree after a loss of control; a Florida firefighter drowned when a tire on his brush truck experienced a blowout and the vehicle rolled over into a water-filled ditch; and a California fire officer died when he was involved in a crash in his personal vehicle while returning from a wildland fire.

Ten firefighters died of heart attacks in association with wildland fires: three heart attacks occurred while on the scene of the fire; two occurred while responding; two occurred in association with pack tests (work capacity tests) taken to certify for wildland firefighting activities; two firefighters experienced heart attacks at home shortly after fighting a wildland fire; and one firefighter suffered a heart attack during nonemergency brush clearing operations.

Three firefighters were killed in wildland aircraft crashes in separate incidents: a helicopter ferrying supplies to fire crews contacted trees in a landing zone and crashed in Washington; a Single Engine Air Tanker (SEAT) crashed while fighting a wildland fire in Utah; and a SEAT crashed during training in Arizona. Many wildland firefighting aircraft were grounded in 2004 in part as a reaction to the numerous wildland firefighting aircraft crashes in 2003.

A California firefighter was killed when her position was burned over in a fast-developing fire, an Illinois firefighter was killed in Arkansas when he was struck by a vehicle as he and his team returned from wildland firefighting activities in Florida, and a firefighter was killed in California when a portion of a tree fell and mortally injured him.

Table 2. Firefighter Deaths Associated with Wildland Firefighting

Year	Total Number of Deaths	Number of Fatal Incidents	Number of Firefighters Killed in Multiple-Death Incidents
2004	21	21	0
2003	29	21	10
2002	23	14	13
2001	15	9	9
2000	17	14	6
1999	26	25	2
1998	14	13	2
1997	12	10	4
1996	9	9	0
1995	18	14	7
1994	36	18	22

Table 3. Wildland Firefighting Aircraft Deaths

Year	Total Number of Deaths	Number of Fatal Incidents
2004	3	3
2003	7	4
2002	6	3
2001	6	3
2000	6	5
1999	0	0
1998	3	2
1997	5	3
1996	0	0
1995	3	1
1994	8	3

TYPE OF DUTY

Activities related to emergency incidents resulted in the deaths of 80 firefighters (Figure 5). This includes all firefighters who died while responding to an emergency, while at an emergency scene, while returning from the emergency incident, and firefighters that died after an incident where the firefighter complained of feeling ill during the incident. Nonemergency activities accounted for 37 fatalities. Nonemergency duties include training, administrative activities, or performing other functions that are not related to an emergency incident. A multiyear historical perspective concerning the percentage of firefighter deaths that occurred during emergency duty is presented in Table 4.

The number of deaths by type of duty being performed in 2004 is shown in Table 5 and presented graphically in Figure 6. Fireground duties returned as the most common type of duty for firefighters killed while onduty. In 2003, an exception occurred in which more firefighters died while responding and returning than for any other type of duty. Unlike the past several years, there were no multiple-firefighter deaths in vehicle crashes in 2004.

Table 5. 2004 Firefighter Deaths by Type of Duty

Type of Duty	Number of Deaths
Fireground Operations	30
Responding/Returning	22
Other Onduty	18
Training	13
Non-fire Emergencies	11
After an Incident	23
Total	**117**

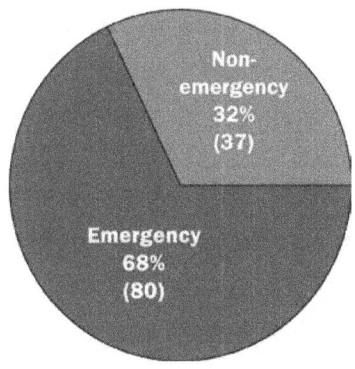

Figure 5. Firefighter Deaths by Type of Duty (2004)

Table 4. Emergency Duty Firefighter Deaths

Year	Percentage of All Deaths
2004	68
2003	71
2002	73
2001	65
2001 w/WTC	92
2000	71
1999	87
1998	77
1997	81
1996	72
1995	86
1994	84

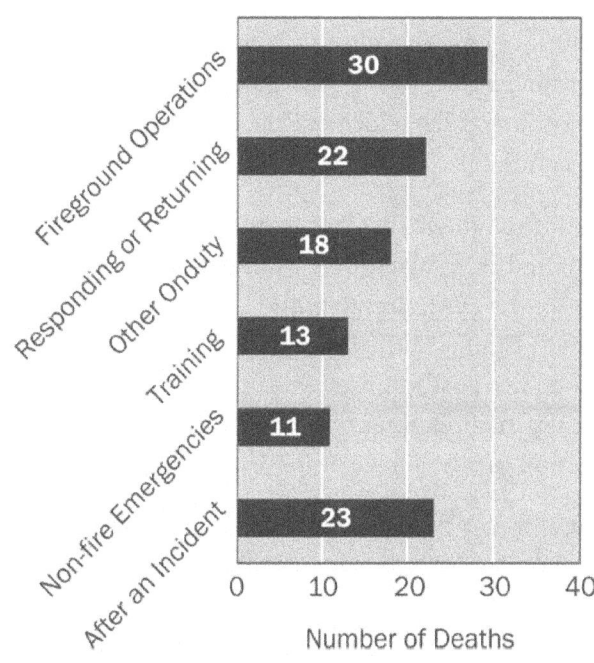

Figure 6. Fatalities by Type of Duty (2004)

Fireground Operations

Thirty firefighters died while engaging in activities at the scene of a fire in 2004. All three of the multiple-firefighter fatality incidents that occurred in 2004 were during structure fires. Two structure fires, in Pittsburgh, Pennsylvania, and in Wood River, Nebraska, claimed two firefighters each as the result of structural collapse, and two firefighters were killed in a basement fire in Philadelphia when one firefighter became entangled in debris.

In addition to the 6 firefighters killed in the multiple-firefighter fatality structural fires described above, 16 firefighters died while engaged in operations at the scene of structural fires: 7 firefighters suffered heart attacks at structure fires; 3 firefighters died after becoming trapped in structural fires in Missouri and Texas; 2 firefighters were killed in fire apparatus backing incidents in New York and California; a Pennsylvania firefighter died as the result of a head injury received as he pulled hose from the rear of a pumper; a Philadelphia fire officer died after partially falling through a floor during a fire in a residential occupancy; a Tennessee fire chief died as a result of injuries he received in a wall collapse at a church fire; and a New Jersey fire officer died in a gas explosion as he helped evacuate residents during a structure fire.

Four firefighters died as the result of heart attacks suffered at wildland fires. A firefighter died in the crash of a SEAT aircraft while fighting a wildland fire in Utah; a California wildland firefighter died when her position was overrun by fire; and a California firefighter was killed when a burning section of a tree fell causing fatal injuries.

A California firefighter was killed as he prepared to fight a vehicle fire at roadside. He was struck by a passing vehicle and thrown on top of fire apparatus on the scene. He was pronounced dead at the scene.

Responding/Returning

Twenty-two firefighters died while responding to or returning from emergency incidents in 2004. Seventeen firefighters died while responding to an emergency incident, and five firefighters died while returning from an emergency. This is a significant drop from the 36 responding and returning deaths reported in 2003 (Table 6).

Vehicle crashes claimed 11 lives while responding; five vehicle-related deaths came in personal vehicle crashes; five firefighters died in apparatus crashes; and a Florida firefighter died in the crash of a medical helicopter while responding to an incident. In one of the response deaths, a Massachusetts firefighter died in a fall from a pumper as it responded to a report of a gas odor in a structure. This same fire department suffered a firefighter fatality as a result of a fall from a similar piece of apparatus in 1984.

In 2004, a firefighter returning from an emergency incident had a blood alcohol level that indicated that the firefighter was intoxicated.

Four firefighters suffered heart attacks while responding, two in the fire station as they prepared for response and two while riding in fire apparatus during a response. One firefighter suffered a thoracic aneurysm, the failure of a blood vessel, while responding and later died.

Four firefighters suffered heart attacks while returning from an incident, and an Illinois firefighter was struck by a truck as he crossed a highway while returning from a wildland incident in Florida.

Table 6. Firefighter Deaths While Responding to or Returning From an Incident

Year	Number of Firefighter Deaths
2004	22
2003	36
2002	13
2001	23
2000	19
1999	26
1998	14
1997	21
1996	22
1995	29
1994	22

Other Onduty

A total of 18 firefighters whose deaths were not associated with the response to any particular emergency died while onduty.

A total of 11 firefighters suffered heart attacks and CVA's while not assigned to an emergency incident or training. Seven firefighters suffered heart attacks or CVA's while onduty in the fire station, while onduty at the fire department administrative offices, or while attending fire department meetings; two firefighters suffered heart attacks while on standby duty, not assigned to an incident; a Florida firefighter suffered a heart attack as he drove a fire department pickup laden with donated equipment; and an Arizona firefighter suffered a heart attack as he participated in a fire department-sponsored brush clearing program.

Two firefighters died in vehicle crashes that were not related to an emergency: a Florida firefighter died as the result of a tire blowout while driving to a controlled (prescribed) burn, and a Pennsylvania firefighter died when the vehicle in which he was a passenger was struck as he and another firefighter ran an errand while onduty.

An Ohio firefighter suffered a fatal head injury after falling from a pickup truck during the cleanup from a fire department fundraiser; a Washington helicopter pilot died when his aircraft contacted trees as he dropped supplies at a landing zone; an Alabama firefighter died when a tree crushed his personal vehicle as he assisted with fallen tree removal in the wake of Hurricane Ivan; an Indiana firefighter died of injuries received when he fell from a horse at a charity event; and a Pennsylvania firefighter died of complications of a bloodborne infection contracted as he worked in floodwaters in the aftermath of heavy rains.

Training

Thirteen firefighters died while engaged in training in 2004.

Heart attacks or heart-related illnesses took the lives of six firefighters engaged in training: two firefighters died during wildland pack (physical capacity) tests; two firefighters died during physical fitness training; and two firefighters died during firefighting training exercises and class work.

A New Hampshire fire officer died while training on SCUBA equipment when he failed to surface in a local lake; a pilot died in the crash of a SEAT aircraft during training in Arizona; a North Carolina firefighter died in a personal vehicle crash en route to training after stopping by the fire station to pick up his equipment; a Tennessee firefighter fell from a pickup and suffered a fatal head injury while riding in the bed of the truck from the fire station to a training site; an Alabama fire chief died of a medication overdose while attending a conference in Florida; a Pennsylvania firefighter was run over and crushed by a fire pumper that he had been photographing for training purposes; and a Kentucky fire chief was killed when he was crushed by a pumper during training on the installation of snow chains on apparatus.

Table 7 offers a multiyear perspective on training deaths.

Table 7. Firefighter Deaths During Training

Year	Number of Firefighter Deaths
2004	13
2003	12
2002	11
2001	14
2000	13
1999	3
1998	12
1997	5
1996	6
1995	3
1994	7

Non-fire Emergencies

Eleven firefighters died at non-fire emergencies. Two firefighters were killed when they were struck by passing vehicles at the emergency scene: a Colorado firefighter was struck as he directed traffic at the scene of a motor vehicle crash, and an Illinois firefighter was killed as he was assisting the driver of a pumper as the apparatus backed onto a street while turning around.

A Kentucky fire officer was killed when she was struck by gunfire on the scene of a domestic violence-related incident. A Florida firefighter was killed in the crash of an ambulance when the driver lost control.

The seven other firefighter deaths related to non-fire emergencies were heart attacks: four occurred at vehicle crashes; two occurred at emergency medical incidents; and a fire police officer suffered a heart attack at a carbon monoxide alarm incident.

After the Incident

Twenty-three firefighters suffered heart attacks after the conclusion of their work shift. Seven of these deaths occurred within 24 hours of an onduty response or training activity and are included in this report as a result of the report criteria change associated with the Hometown Heroes Survivors Benefit Act of 2003. This fact does not diminish the tragedy or importance of these deaths.

Nineteen of the firefighter deaths that occurred after the incident were heart attacks and two were CVA's. A Pennsylvania firefighter died as a result of a ruptured aorta after an incident response, and a California fire officer died in a vehicle crash after completing work at a wildland fire.

In six of these incidents, firefighters specifically complained of not feeling well while onduty. In several cases, these firefighters told other firefighters that they were going home to rest and instead went home and suffered fatal heart attacks.

Career, Volunteer, and Wildland Deaths by Type of Duty

Figure 7 depicts career, volunteer, and wildland firefighter deaths by type of duty. Wildland career, wildland seasonal, and wildland contractor deaths were grouped together. As in past years, there were a disproportionate number of fatalities experienced by volunteer firefighters responding to and returning from alarms as compared to career firefighters. Fourteen volunteer firefighter deaths occurred while responding, and three occurred while returning from an emergency. Of the responding deaths, eight were due to vehicle crashes.

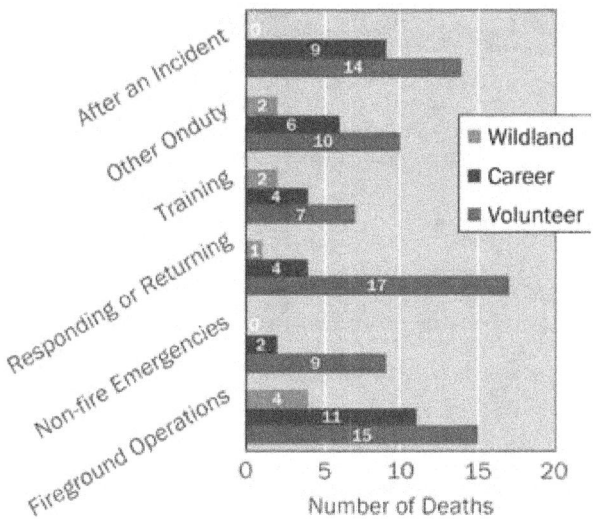

Figure 7. Career, Volunteer, and Wildland Deaths by Type of Duty (2004)

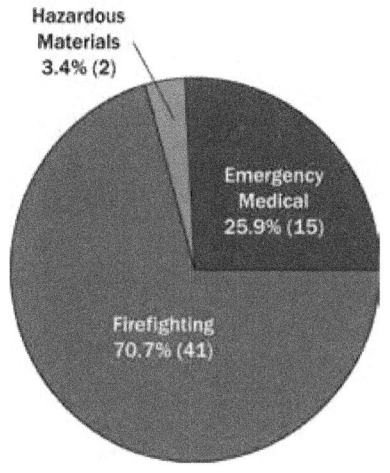

Figure 8. Type of Emergency Duty

Type of Emergency Duty

In 2004, 58 firefighters died while engaged directly in the delivery of emergency services. This number includes deaths that were the result of injuries sustained on the incident scene or en route to the incident scene.

Figure 8 shows the number of firefighters killed in firefighting, emergency medical services, technical rescue-related incidents, and other emergency incidents in 2004.

An additional 22 firefighter fatalities occurred in association with emergencies but not while firefighters were engaged in the direct delivery of emergency services on-scene (returning from the emergency and after the emergency).

Forty-one firefighters were killed in relation to fires; 15 firefighters were killed in relation to EMS calls; and 2 firefighters were killed at emergencies that involved hazardous materials.

CAUSE OF FATAL INJURY

The term *cause of injury* refers to the action, lack of action, or circumstances that resulted directly in the fatal injury. The term *nature of injury* refers to the medical cause of the fatal injury or illness, and often is referred to as the physiological cause of death. A fatal injury is usually the result of a chain of events, the first of which is recorded as the cause.

Table 8 and Figure 9 show the distribution of deaths by cause of fatal injury or illness.

Table 8. Cause of Fatal Injury – 2004

Cause	Number
Stress/Overexertion	66
Vehicle Collision	20
Struck by	10
Caught/Trapped	8
Collapse	6
Fall	5
Other	2
Total	**117**

2004 had more stress-/overexertion-related deaths than any year in over a decade.

15

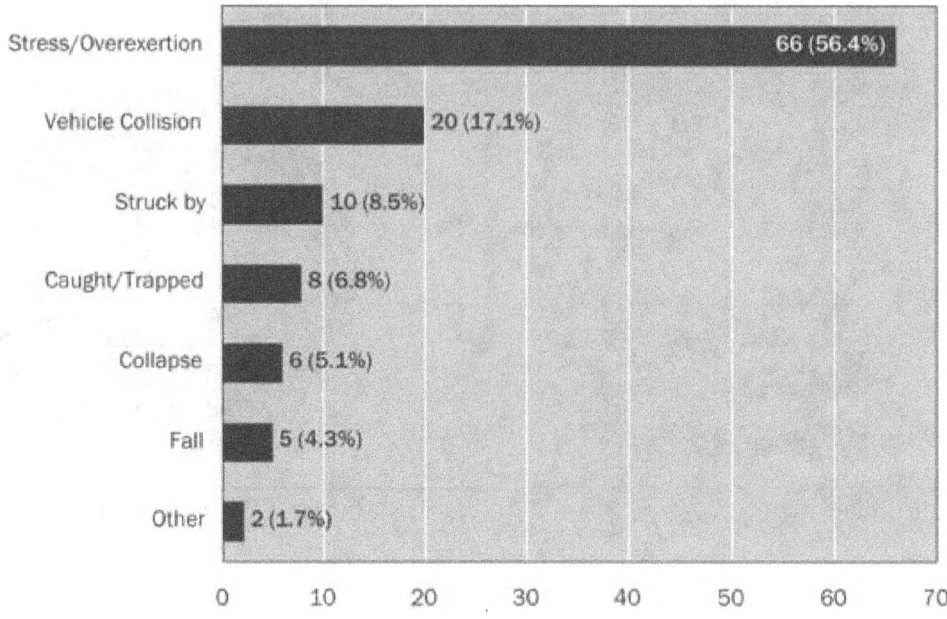

Figure 9. Fatalities by Cause of Fatal Injury (2004)

Stress or Overexertion

Stress or overexertion is a general category that includes all firefighter deaths that are cardiac or cerebrovascular in nature, such as heart attacks and strokes (CVA's), and other illnesses, such as extreme climatic heat exposure. Classification of a firefighter fatality in this cause of fatal injury category does not indicate that a firefighter was in poor physical condition.

Firefighting is extremely strenuous physical work and is likely one of the most physically demanding activities that the human body performs. As it has been in every year for more than the past decade, the largest cause of firefighter deaths in 2004 was stress or overexertion. Stress or overexertion was listed as the primary factor in 66 firefighter deaths in 2004. This is the highest number of stress-related deaths in a decade. Even if the firefighters included in this year's study as a result of the inclusion criteria change are held aside momentarily, 51.9 percent of the remaining firefighter deaths in 2004 were due to stress or overexertion, the highest percentage in at least a decade.

Most firefighter deaths attributed to stress result from heart attacks. Of the 66 stress-related fatalities in 2004, 61 firefighters died of heart attacks, 4 died as a result of CVA's, and 1 died of an aortic aneurysm.

Table 9. Deaths Caused by Stress or Overexertion

Year	Number	Percent of Fatalities
2004	66	56.4
2003	51	45.9
2002	38	38.0
2001	43	40.9*
2000	46	44.6
1999	56	49.5
1998	43	46.2
1997	41	41.0
1996	46	46.4
1995	49	48.0
1994	36	34.2

* Does not include the firefighter deaths of September 11, 2001, in New York City.

16

Vehicle Crashes

As in most years, the second leading cause of fatal injury for firefighters who died in 2004 was vehicle crashes. Unlike the most recent past, however, there were no multiple-firefighter fatality incidents involving vehicle crashes. A total of 20 firefighters died in 2004 in vehicle crashes.

The reduction in number to 20 firefighters killed in vehicle crashes in 2004 is a welcome change to the pattern of the last 5 years where deaths due to vehicle-related incidents had been rising each year (see Figure 10). Although this lower number is good news, the number of vehicle-related deaths is still unacceptably high. Most of these deaths could have been prevented by lowering response speeds and the use of safety belts.

In 2004, eight firefighters died in personally owned vehicle (POV) crashes, four died in vehicle-related incidents involving engine or pumper apparatus, four firefighters died in aircraft crashes, two firefighters died in crashes involving brush firefighting apparatus, and one firefighter died in

Of the 16 nonaircraft-related vehicle crash firefighter deaths in 2004, safety belts or seatbelts were only known to be in use in 5 instances. The continued lack of seatbelt use in the fire service has led to needless firefighter deaths each year.

each of two crashes that involved an ambulance and a tanker (tender).

Deaths in POV's continue to be a major concern. In 2004, eight firefighters lost their lives in POV crashes. Seven of these firefighters were volunteer members of their departments, and one was a career firefighter. Only one of the volunteer firefighters killed in a POV crash in 2004 was over the age of 20. Only one of the eight firefighters killed in a POV crash in 2004 was known to be wearing a seatbelt.

Four firefighters died in 2004 in vehicle incidents that involved engines or pumpers: an Illinois firefighter died when he was ejected from a pumper responding to a structure fire after it

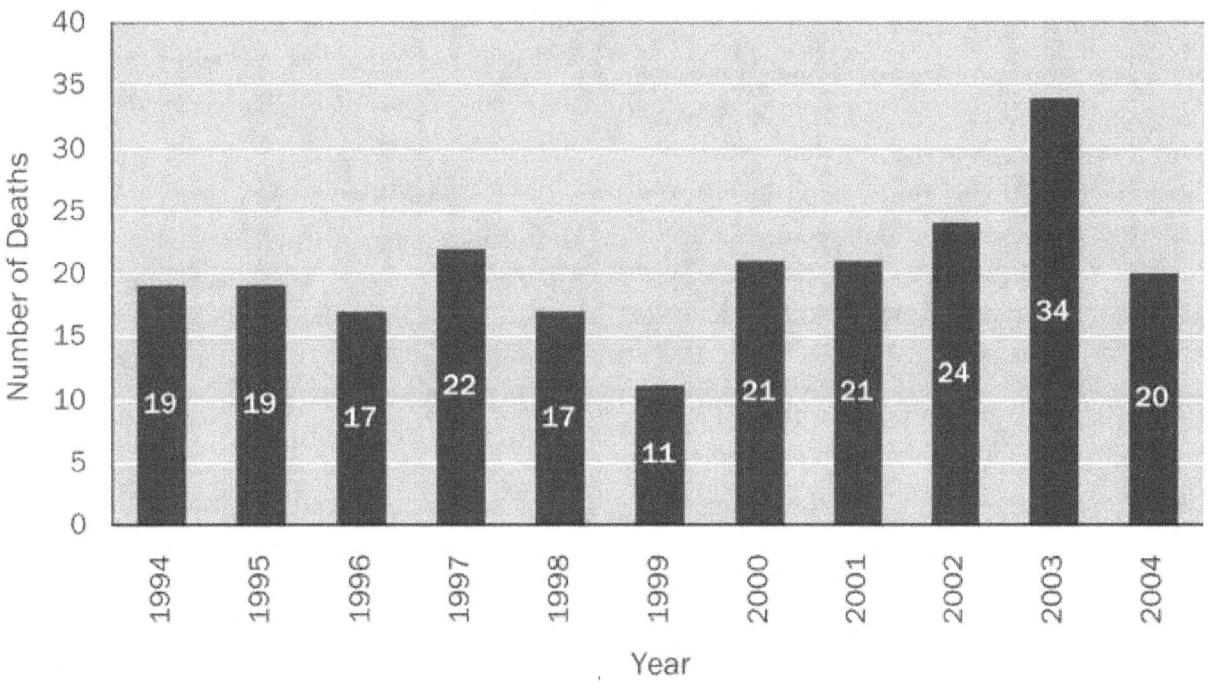

Figure 10. Firefighter Fatalities in Vehicle Crashes

became involved in a collision with another pumper responding to the same incident; a California firefighter was killed when she somehow left her position on the back step of a pumper as it backed out of an incident scene and was crushed under the wheels of the apparatus; a Missouri fire officer was killed when his apparatus was involved in a collision, left the roadway, and crashed into a tree; and a Kentucky firefighter was ejected from a responding engine apparatus after she lost control of the vehicle and left the roadway.

Aircraft crashes claimed four firefighters in 2004: a Florida firefighter was killed in the crash of a medical helicopter as it responded to an incident in poor weather; a SEAT pilot died as a result of a crash during training; a SEAT pilot died after a crash while fighting a wildland fire in Utah; and a Washington helicopter pilot died when his tail rotor contacted trees and the aircraft crashed while dropping supplies at a landing zone.

Two firefighters died in crashes involving brush firefighting vehicles: a Florida firefighter drowned after his vehicle experienced a tire blowout and ended up on its roof in a water-filled ditch; and a Mississippi firefighter died in the crash of a brush/rescue vehicle (she was trapped in the wreckage, and extrication equipment was inaccessible due to damage to the vehicle).

A Mississippi firefighter was killed after he lost control of a fire tanker while avoiding another vehicle in a curve, and a Florida firefighter was killed while providing patient treatment in the rear of an ambulance that was transporting a customer to the hospital.

A brush truck driven by Firefighter Edward Orlando Peters rolled over into a water-filled ditch as a result of a tire blowout.

PHOTO BY FLORIDA DIVISION OF FORESTRY

Struck by Object

Being struck by an object was the third leading cause of fatal firefighter injuries in 2004. Ten firefighters died after being struck by vehicles or other objects while onduty.

Three firefighters were struck by passing vehicles as they worked at emergency scenes: a California firefighter was killed as he prepared to fight a vehicle and grass fire; a Colorado firefighter was killed as he directed traffic at the scene of an MVA; and an Illinois firefighter was killed as he acted as the spotter for an engine apparatus as it backed out of a driveway.

Two firefighters were killed by falling trees: a California wildland firefighter was killed when a portion of a tree fell and struck him, and an Alabama firefighter was killed when a storm-damaged tree fell onto his pickup truck as he worked to clear tree debris from roadways after the passage of a hurricane.

Two firefighters were killed when they were struck by fire apparatus, one at the scene of a structure fire and one at the fire station: a New York firefighter was killed when he was struck by a tanker and pinned between the tanker and a pumper drafting from a portable water tank, and a Pennsylvania firefighter was struck and crushed by a pumper as it backed into the fire station after the firefighter took pictures of the unit for training purposes.

A Kentucky fire chief was killed when he was struck and crushed by a fire truck during a training exercise on installing snow chains; the apparatus rolled forward after an airbag lifting the vehicle dislocated. An Illinois wildland firefighter was killed in Arkansas when he was struck by a passing truck as he crossed the highway; the firefighter and his crew had stopped for the night while returning to Illinois from wildland firefighting in Florida.

Caught or Trapped

In 2004, eight firefighters were killed when they were caught or trapped. This classification covers firefighters who are trapped in wildland and structural fires and unable to escape due to rapid fire progression and byproducts of smoke, heat, and flame. This classification also includes firefighters who are killed by drowning.

Six firefighters were killed in structure fires when they were trapped by fire progress: two Philadelphia firefighters were killed in the basement of a residence when fire conditions worsened rapidly and one firefighter became entangled in debris (the firefighter's company officer stayed and attempted to free the entangled firefighter, but both were killed); two firefighters died after becoming trapped by fire progress in drinking establishments, one in Missouri and one in Texas; a Texas firefighter became trapped in a structure fire in a residence; and a New Jersey chief officer was killed when propane cylinders in the common hallway of a multiple residence vented their contents during a fire and caused an explosion.

A California wildland firefighter was killed when a wind shift caused a fire to grow rapidly and made a run toward firefighters; she attempted to run to an established safety zone but was not able to reach it in time.

A New Hampshire fire officer drowned as he and another firefighter trained in SCUBA gear in the open water portion of a frozen lake. His body was recovered the next day.

Collapse

Six firefighters were killed in collapses in 2004. Two multiple-fatality incidents took the lives of two firefighters each: two Wood River, Nebraska, firefighters were killed in the collapse of an ice-covered roof during a fire in a residence; and a Pittsburgh, Pennsylvania, battalion chief and firefighter were killed in the sudden collapse of a church bell tower several hours into a major fire in a historic downtown church.

A Philadelphia fire officer was killed during a structural fire in a residence when he fell partially through a fire weakened floor and was trapped. A Tennessee fire chief was killed when a wall collapsed during a fire in a church; the chief died almost four months after he was injured.

Firefighters working on the scene of a fire in a historic church in Pittsburgh, Pennsylvania. The bell tower collapsed shortly after this photograph was taken and killed Battalion Chief Charles G. Brace and Master Firefighter Richard A. Stefanakis.

PHOTO BY TONY TYE, PITTSBURGH POST-GAZETTE

Falls

Five firefighters died in 2004 as the result of falls. Two firefighters were killed in separate incidents when they fell from the cargo or bed area of pickup trucks; one firefighter was riding to a training site, and the other was returning supplies used for a fundraising event when they fell (both firefighters suffered fatal head injuries).

A Pennsylvania firefighter fell from the rear step of a pumper and suffered a fatal head injury as he stretched hose at the scene of a structural fire. An Indiana firefighter fell from a horse during a community event and died of a head injury.

A Massachusetts firefighter fell from a responding engine company as it turned out of the fire station while en route to an incident. The door of the apparatus opened, and the firefighter received a fatal head injury when he struck the pavement. There were reports that the firefighter was not wearing a seatbelt and that there had been problems with the latch on the door that opened.

Other

Two firefighters died in 2004 of causes that are not categorized above: a Pennsylvania firefighter died as a result of a bloodborne infection that he likely contracted during extended operations in floodwater after severe weather in the area, and an Alabama fire chief died as a result of an accidental overdose of prescribed medications while attending a conference in Florida.

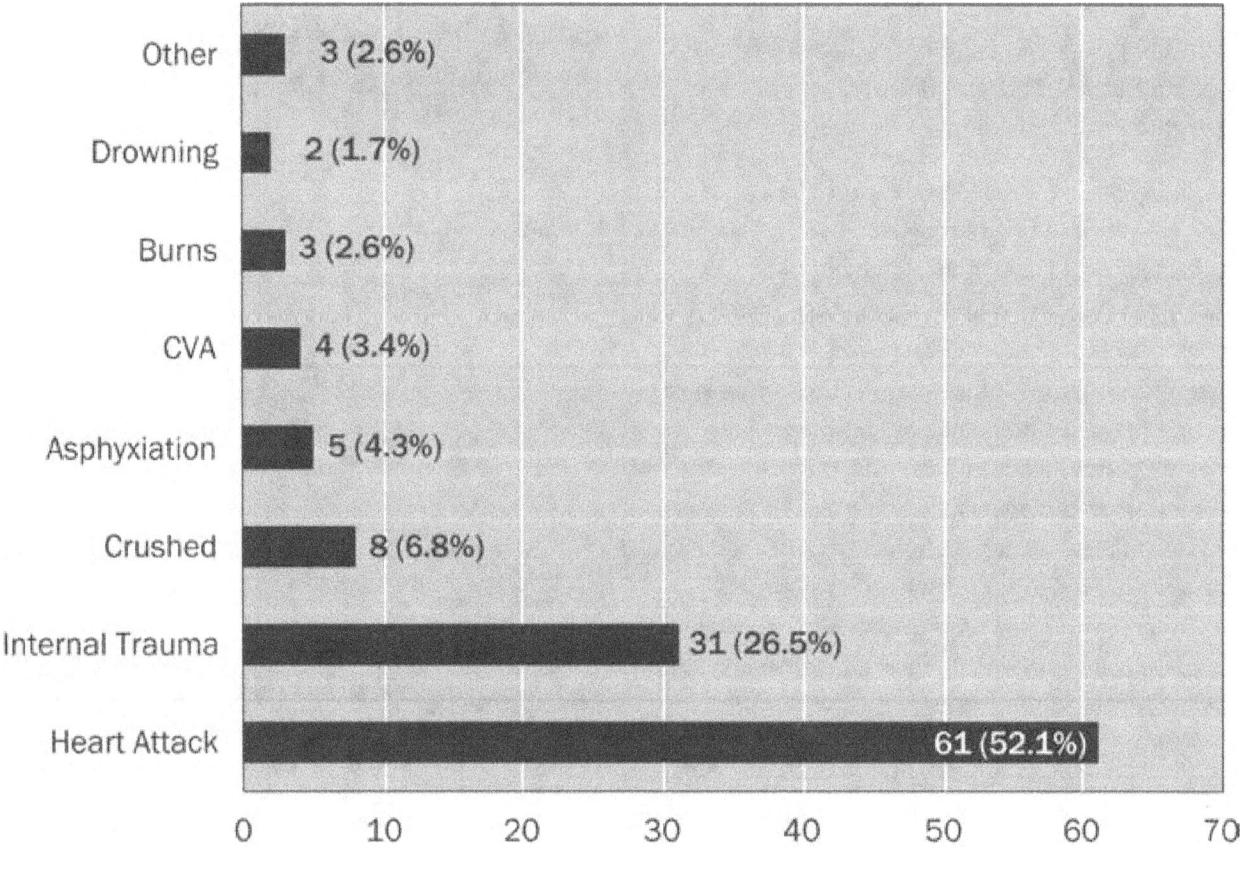

Figure 11. Fatalities by Nature of Fatal Injury (2004)

NATURE OF FATAL INJURY

Table 11 and Figure 11 show the distribution of the 117 deaths by the medical nature of the fatal injury or illness.

Table 11. Nature of Fatal Injury – 2004

Nature	Number
Heart Attack	61
Internal Trauma	31
Crushed	8
Asphyxiation	5
CVA	4
Burns	3
Drowning	2
Other	3
Total	**117**

Heart Attack

Again in 2004, heart attacks were the most frequent nature of death with 61 firefighter deaths. Figure 12 provides a detailed breakdown of heart attacks by type of duty.

Nineteen firefighters died of heart attacks suffered after the conclusion of their onduty activities: 11 firefighters suffered heart attacks at home after returning from onduty activities; 3 firefighters suffered heart attacks at the fire station after a response; 3 firefighters were participating in physical fitness activities after going offduty; 1 firefighter was driving his personal vehicle; and 1 firefighter was working at a second job when he became ill.

Eleven firefighters suffered heart attacks while engaged in activities on the fire scene: two of the firefighters who died were fire police officers engaged in traffic control at incident scenes; five

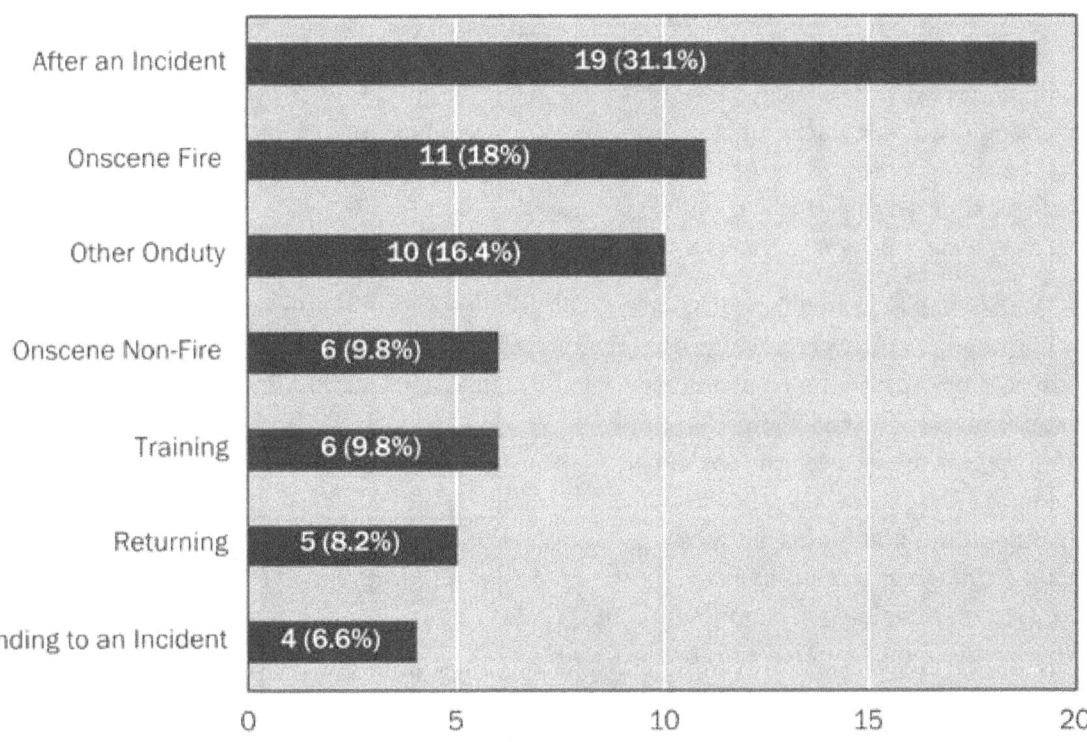

Figure 12 - Heart Attacks by Type of Duty (2004)

firefighters had heart attacks during structure fires; and four firefighters suffered heart attacks at wildland fires.

The median age of the firefighters who died of an onduty heart attack in 2004 was 52.9 years old. The youngest victim was 23 years old, and the oldest was 78 years old.

Ten firefighters died of heart attacks in 2004 while onduty but were not assigned to any specific incident or training exercise: three firefighters became ill while completing regular duties in the fire station; two firefighters suffered heart attacks while attending meetings in the fire station; one firefighter suffered a heart attack while providing standby services; a Florida firefighter became ill while transporting donated equipment; a Kansas fire officer became ill while at his desk in an administrative facility; an Arizona firefighter suffered a heart attack and later died while participating in a fire department-sponsored brush clearing operation; and a military firefighter suffered a cardiac event while on standby duty.

Six firefighters suffered heart attacks during their response to non-fire emergencies: three firefighters became ill while working at the scene of motor vehicle crashes; two firefighters suffered heart attacks at emergency medical incidents; and one fire police officer suffered a heart attack while directing traffic at a carbon monoxide alarm response.

Six firefighters died of heart attacks that struck during training: two firefighters suffered heart attacks during or shortly after completion a pack (work capacity) test for wildland firefighting certification; two firefighters, one in Colorado and one in Idaho, suffered heart attacks while participating in physical fitness duties; a Connecticut firefighter was struck with a heart attack during maze training at a training facility; and a Florida firefighter suffered a heart attack and died while attending a firefighter certification class at a local community college.

Five firefighters suffered heart attacks while responding to an emergency: two firefighters collapsed in the fire station prior to response to incidents; two firefighters became ill while riding in fire apparatus; and one firefighter became ill while a passenger in a responding POV.

Four firefighters became ill and later died of heart attacks while returning from an emergency: a fire chief suffered a heart attack in his car while preparing to leave the scene of a controlled burn; two firefighters became ill while returning to the fire station on fire apparatus; and a fire chief became ill while driving a vehicle back from a response (the vehicle left the road and came to a stop, and other firefighters came to the chief's side and provided medical assistance).

Internal Trauma

In 2004, 31 firefighters died due to internal physical trauma. This grouping includes most firefighters who are killed in vehicle crashes and those who receive physical injuries.

Table 12. Internal Trauma Firefighter Deaths

Year	Number of Firefighter Deaths
2004	31
2003	41
2002	34
2001	28*
2000	36
1999	25
1998	27
1997	32
1996	32
1995	24
1994	21

* Does not include the firefighter deaths of September 11, 2001, in New York City.

Thirteen traumatic deaths were the result of motor vehicle crashes. Six firefighters were struck by motor vehicles and died of physical trauma. Four firefighters died in aircraft crashes.

Four firefighters died from traumatic wounds suffered in falls from vehicles, and one died from injuries received in a fall from a horse. One firefighter died from traumatic injuries received in a building collapse, one firefighter was injured by a falling section of tree and later died, and a fire officer was struck by gunfire.

Crushed

In 2004, eight firefighters died when they were crushed: two Wood River, Nebraska, firefighters were crushed under the weight of an ice-covered roof that collapsed; two Pittsburgh, Pennsylvania, firefighters died when they were crushed by debris from a fallen bell tower at a major fire in an historic church; two firefighters were crushed by fire apparatus, one on the scene of a structure fire and one at the fire station; one firefighter was crushed in a motor vehicle crash; and an Alabama firefighter was crushed by a fallen tree while driving his pickup truck.

Asphyxiation

Asphyxiation was the fourth leading medical reason for firefighter deaths in 2004, responsible for five deaths: four firefighters died of asphyxiation after becoming trapped in structure fires, two in Philadelphia and one each in Missouri and Texas; and a California firefighter died of smoke inhalation when her position was overrun in a wildland fire. Table 13 provides an historical perspective on asphyxiation deaths.

Cerebrovascular Accident (CVA)

Four firefighters died in 2004 as a result of strokes (CVA's): two firefighters complained of not feeling well during an incident and then sought treatment after going home; a firefighter did not feel well at the scene of an incident, then suffered a CVA while driving his personal vehicle in the fire

station parking lot; and a firefighter suffered a CVA while onduty in the fire station.

Table 13. Firefighter Deaths due to Asphyxiation

Year	Number of Firefighter Deaths
2004	5
2003	6
2002	15
2001	18
2000	13
1999	16
1998	15
1997	15
1996	5
1995	20
1994	29

Burns

Three firefighters died as a result of burns in 2004. All of these injuries occurred during structural firefights: a Texas firefighter died of burns received in an arson-caused fire in a nightclub; a New Jersey chief officer died of burns received in a multiple residence fire; and a Philadelphia fire officer received burns in a partial floor collapse and entrapment.

Drowning

Two firefighters drowned in 2004: a Florida firefighter drowned after a tire blowout sent his brush truck into a water-filled ditch upside down, and a New Hampshire fire officer died during SCUBA training.

Other

Three firefighters died in situations where the nature of their deaths does not fall into any of the categories described above: a Pennsylvania fire officer died of a ruptured aorta; a Pennsylvania fire police officer contracted a bloodborne infection while doing extensive duty in contaminated floodwaters; and an Alabama fire chief died as a result of an unintentional overdose of prescribed medications while attending a business conference in Florida.

FIREFIGHTER AGES

Figure 13 shows the percentage distribution of firefighter deaths by age and nature of the fatal injury. Table 14 provides counts of firefighter fatalities by age and the nature of the fatal injury.

As in most years, younger firefighters were more likely to have died as a result of traumatic injuries such as injuries from an apparatus accident or after becoming caught or trapped during firefighting

Years of Service: Firefighters who died onduty in 2004 had a median of 16 years of service. This compares to 14 years of service for firefighters who died in 2003, 11 years of service for firefighters who died in 2002, and 11.5 years of service for firefighters who died in 2001, not including those who died at the World Trade Center.

operations. Stress plays an increasing role in firefighter deaths as age increases.

Table 14. Firefighter Ages and Nature of Fatal Injury

Age Range	Nontrauma Total	Trauma Total
under 21	0	6
21 to 25	1	5
26 to 30	1	3
31 to 35	1	5
36 to 40	4	8
41 to 45	9	4
46 to 50	15	2
51 to 60	23	10
61 & over	14	6

The youngest firefighter killed in 2004 was Junior Firefighter Joshua Martin of Louisiana, age 15. The oldest firefighter killed was Fire Police Captain Thomas Conway of New Jersey, age 81.

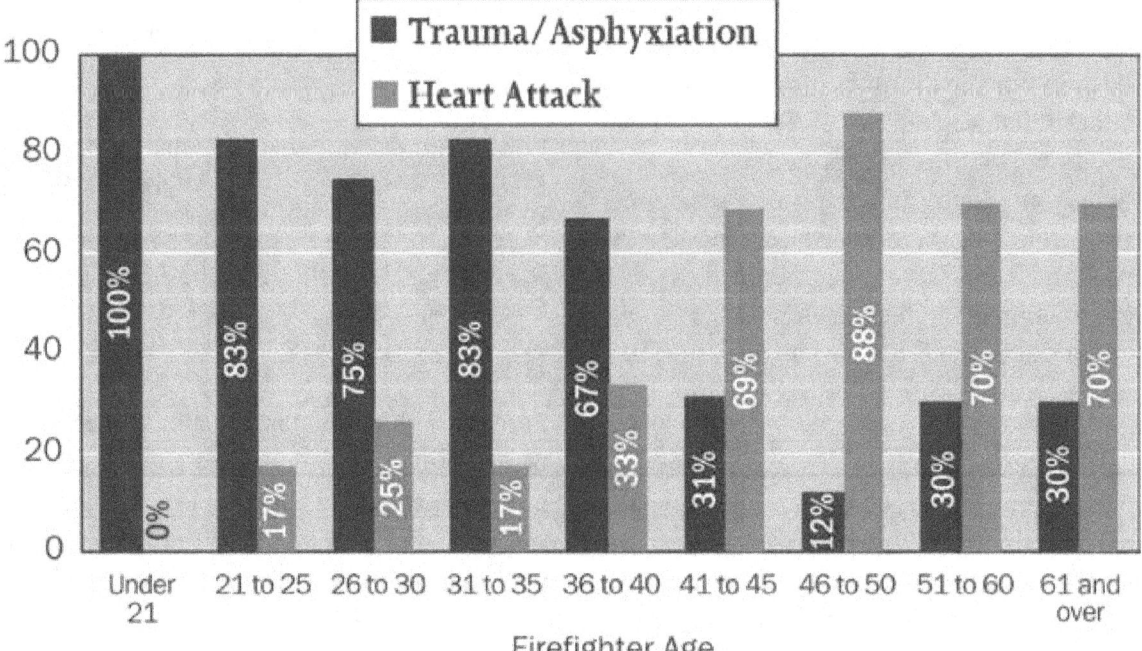

Figure 13. Fatalities by Age and Nature (2004)

FIXED PROPERTY USE FOR STRUCTURAL FIREFIGHTING DEATHS

There were 22 firefighter fatalities in 2004 where the firefighters became ill while on the scene or engaged in structural firefighting. Table 15 shows the distribution of these deaths by fixed property use. As in most years, residential occupancies accounted for the highest number of these fireground fatalities, with 15 deaths.

Table 16 shows the number of firefighter deaths in residential occupancies for the past 11 years. Residential occupancies usually account for 70 percent to 80 percent of all structure fires and a similar percentage of the civilian fire deaths each year*. Historically, the frequency of firefighter deaths in relation to the number of fires is much higher for nonresidential structures.

Table 15. Structural Firefighting Deaths by Fixed Property Use in 2003

Fixed Property Use	Number	Percent
Residential	15	68
Commercial	7	32

Table 16. Firefighter Deaths In Residential Occupancies

Year	Number of Firefighter Deaths
2004	15
2003	10
2002	21
2001	17
2000	21
1999	23
1998	17
1997	16
1996	19
1995	18
1994	25

TYPE OF ACTIVITY

Table 17 and Figure 14 show the types of fireground activities in which firefighters were engaged at the time they sustained their fatal injuries or illnesses. This total includes all firefighting duties such as wildland firefighting and structural firefighting. In 2004 there were a total of 30 firefighter deaths on the fireground.

Table 17. Type of Activity – 2004

Nature	Number
Fire Attack	16
Search and Rescue	4
Water Supply	3
Scene Safety	2
Incident Command	2
Ventilation	2
Suppression Support	1
Total	30

Fire Attack

In 2004, 16 firefighters were killed as they engaged in direct fire attack, such as advancing or operating a hoseline at a fire scene. In years past, most fireground firefighter deaths occurred while the firefighter was engaged in fire attack (see Table 18).

One multiple-firefighter fatality incident took the lives of two Philadelphia firefighters engaged in fire attack in a residential basement. A Philadelphia fire officer died in a residential fire earlier in the year when he became caught in a structural collapse.

Four firefighters died in the interior of fire-involved structures while engaged in fire attack duties: a Texas firefighter died in a residential fire; another Texas firefighter died in a fire in a commercial building; a Missouri firefighter died

---

* Complete 2004 NFIRS fire incidence data were not available at the time of this report, but residential fires typically account for between 70 percent and 80 percent of all civilian fatalities each year according to the NFPA.

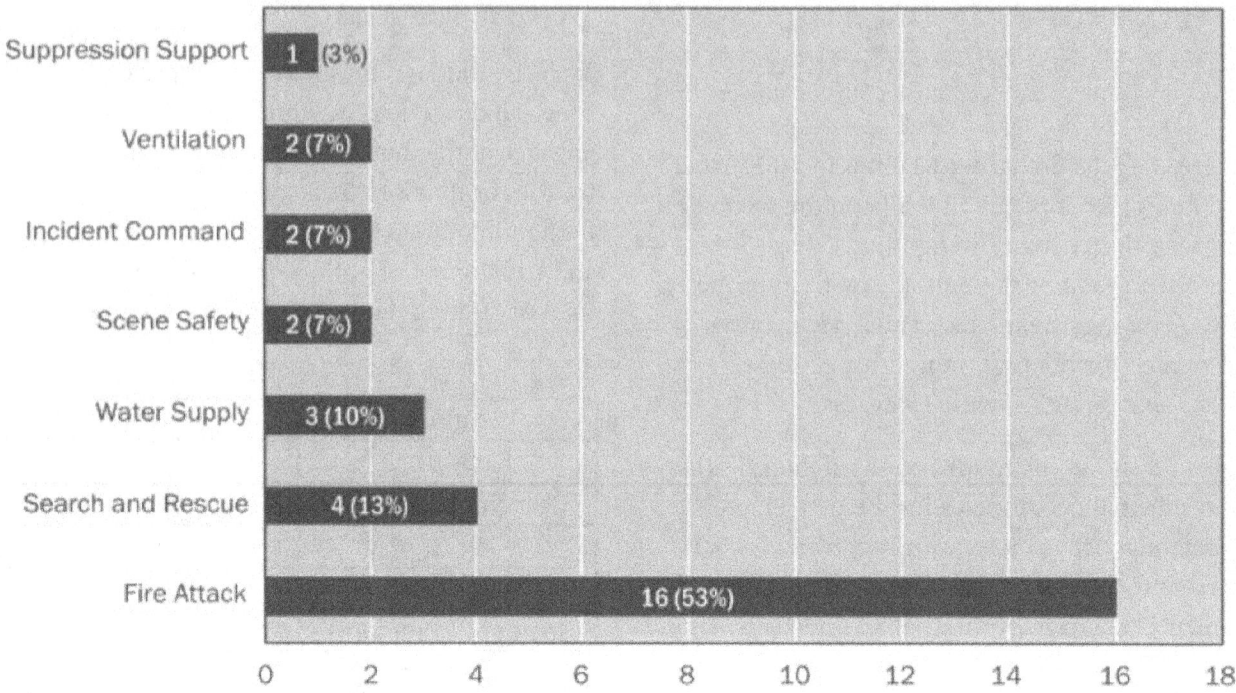

Figure 14. Fatalities by Type of Activity (2004)

in a commercial building; and a Pennsylvania firefighter died in the collapse of a historic church.

Four firefighters suffered heart attacks while engaged in fire attack duties: a firefighter in New York and a firefighter in Illinois died while advancing hoselines on the exterior of a fire involved structure; a Kentucky firefighter became ill at the scene of a fire in a manufactured home; and a South Carolina firefighter became ill at the scene of a fire in a community center.

Four firefighters died while engaged in fire attack at wildland incidents: a SEAT pilot crashed after a fire retardant drop at a fire in Utah; a firefighter in California was killed when fire overran her position as she and other firefighters attacked a fire; and firefighters in Alabama and Texas suffered heart attacks at wildland fires.

A California firefighter was struck and killed by a passing vehicle as he prepared to advance a hoseline on a car and grass fire.

The body of Firefighter Eva Marie Schicke is placed into her crew's helicopter by members of her helitack crew. She was killed when fire overran her position as she and other firefighters attacked a wildfire in California.

PHOTO BY AL GOLUB, THE MODESTO BEE

Table 18. Firefighter Deaths While Engaged in Fire Attack

Year	Number of Firefighter Deaths
2004	16
2003	11
2002	13
2001	13
2000	13
1999	16
1998	18
1997	21
1996	9
1995	18
1994	7

Search and Rescue

Four firefighters were killed in 2004 as they engaged in search and rescue activities: two Wood River, Nebraska, firefighters were killed as they searched a residence for a fire victim when the ice-covered roof of an addition to the home collapsed; an Ohio firefighter became ill after completing search and rescue activities at a structural fire and later suffered a heart attack; and a New Jersey chief officer was killed as he notified building residents of a fire when he was engulfed in a fireball resulting from venting propane cylinders.

Water Supply

Three firefighters died in 2004 while engaged in water supply duties: a New York firefighter was killed when he was crushed between a pumper and a tanker at a structure fire; a Pennsylvania firefighter died after receiving a severe head injury when he fell while removing supply line from the rear of a pumper at a structure fire; and a North Dakota firefighter died of a heart attack suffered while operating the pump on a tanker (tender).

Scene Safety

A Pennsylvania fire police officer suffered a heart attack while directing traffic at a structural fire in an industrial occupancy, and a Mississippi-based firefighter suffered a heart attack in Florida during the performance of safety duties at a wildland fire.

Incident Command

Two Incident Command officers were killed in 2004, both in church structural collapses: a Pittsburgh, Pennsylvania, chief officer was killed in the collapse of a historic church; and a Tennessee fire chief was killed when the front wall of a church collapsed and struck him.

Ventilation

Two firefighters died in 2004 while engaged in ventilation duties, both at fires in residential occupancies.

Suppression Support

One firefighter was killed in 2004 while engaged in suppression support. A California firefighter was killed when he was struck by a falling portion of a tree as he prepared to cut down dangerous trees at a wildland fire.

A ladder arch awaits the arrival of the funeral procession of Firefighter Jackson Huber Gerhart after his funeral.

PHOTO BY CHARLES ARMSTRONG, THE PUBLIC OPINION

TIME OF INJURY

Figure 15 illustrates the distribution of all 2004 firefighter deaths according to the time of day when the fatal injury occurred. The time of fatal injury for six firefighters either was not known or was not reported.

MONTH OF YEAR

Figure 16 illustrates firefighter fatalities by month of the year.

STATE AND REGION

Table 19 shows the distribution of firefighter deaths by State. Firefighters based in 41 States died in 2004.

The highest number of firefighter deaths based on the location of the fire service organization in 2004 occurred in Pennsylvania with 18 deaths.

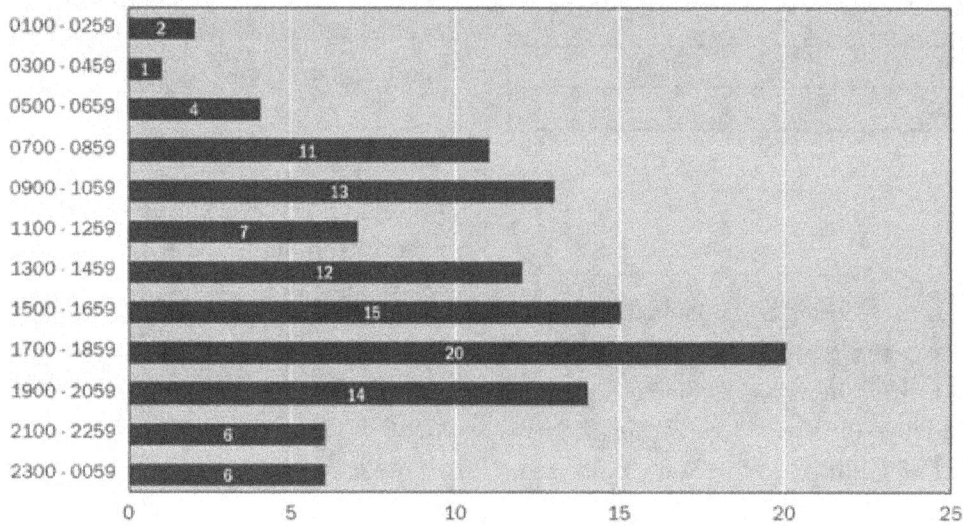

Figure 15. Fatalities by T me of Fatal Injury (2004)

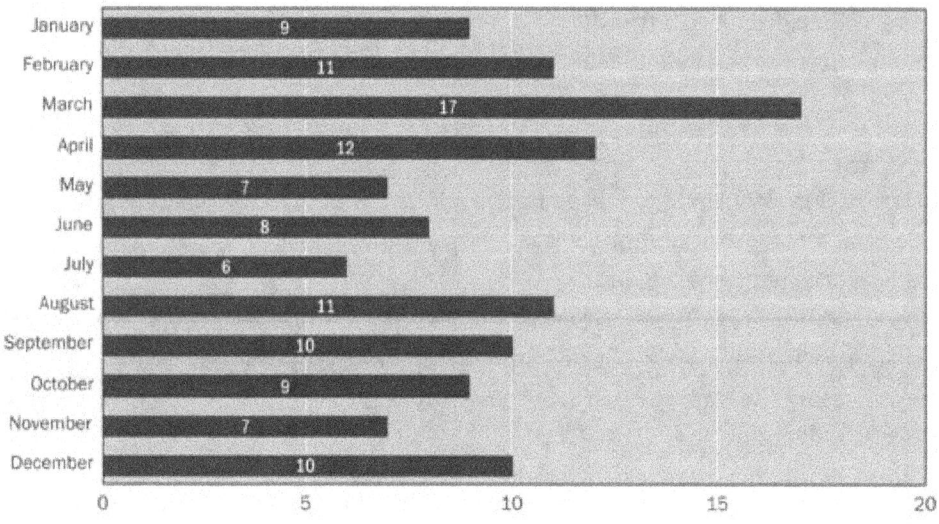

Figure 16. Deaths by M nth of the Year (2004)

Table 19. Firefighter Fatalities by State by Location of Fire Service*

Number	State	Percent of Total	Number	State	Percent of Total
1	Alaska	0.85%	3	Mississippi	2.56%
3	Alabama	2.56%	1	Montana	0.85%
1	Arkansas	0.85%	3	North Carolina	2.56%
1	Arizona	0.85%	1	North Dakota	0.85%
5	California	4.27%	2	Nebraska	1.70%
2	Colorado	1.70%	1	New Hampshire	0.85%
1	Connecticut	0.85%	6	New Jersey	5.12%
5	Florida	4.27%	1	New Mexico	0.85%
3	Georgia	2.56%	4	New York	3.41%
2	Iowa	1.70%	3	Ohio	2.56%
2	Idaho	1.70%	1	Oklahoma	0.85%
7	Illinois	5.98%	1	Oregon	0.85%
1	Indiana	0.85%	18	Pennsylvania	15.3%
3	Kansas	2.56%	1	Rhode Island	0.85%
8	Kentucky	6.83%	4	South Carolina	3.41%
1	Louisiana	0.85%	2	Tennessee	1.70%
2	Massachusetts	1.70%	4	Texas	3.41%
2	Maryland	1.70%	1	Utah	0.85%
1	Michigan	0.85%	2	Virginia	1.70%
4	Missouri	3.41%	1	Washington	0.85%
			2	Wisconsin	1.70%

* This list attributes the deaths according to the State in which the fire department or unit is based, as opposed to the State in which the death occurred. They are listed by those States for statistical purposes and for the National Fallen Firefighters Memorial at the National Emergency Training Center.

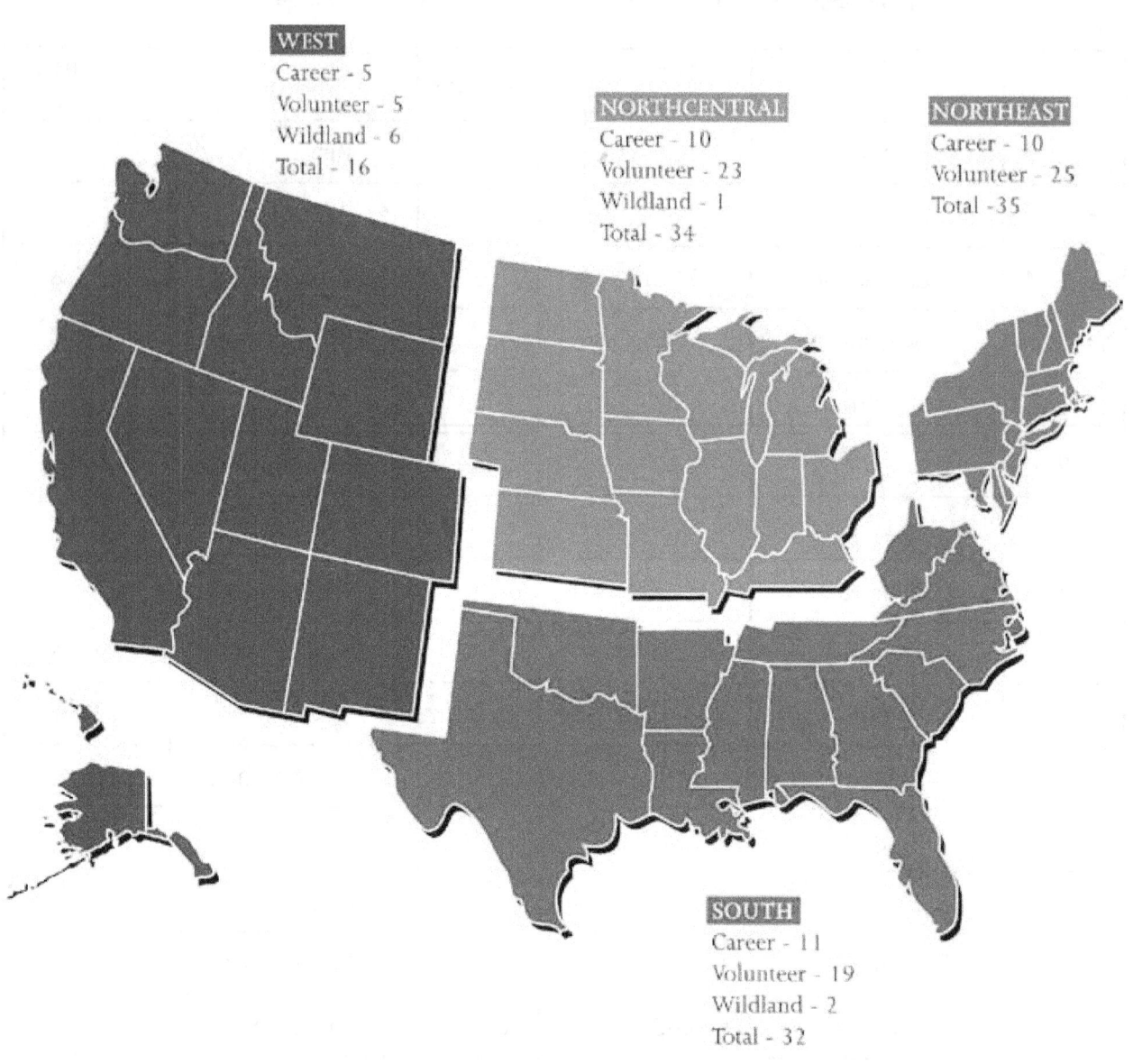

WEST
Career - 5
Volunteer - 5
Wildland - 6
Total - 16

NORTHCENTRAL
Career - 10
Volunteer - 23
Wildland - 1
Total - 34

NORTHEAST
Career - 10
Volunteer - 25
Total - 35

SOUTH
Career - 11
Volunteer - 19
Wildland - 2
Total - 32

Figure 17. Firefighter Fatalities by Region (2004)

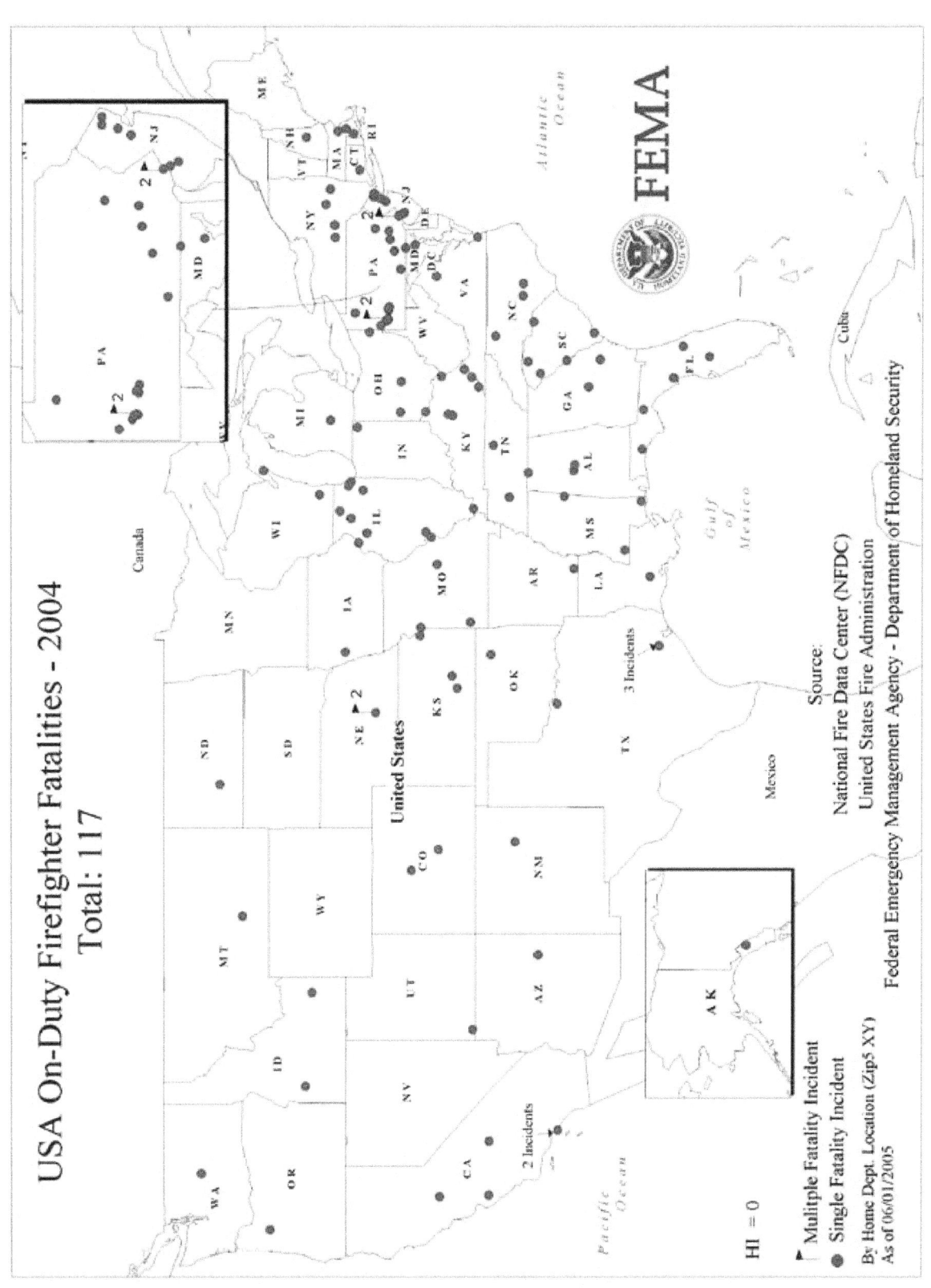

USA On-Duty Firefighter Fatalities - 2004
Total: 117

31

ANALYSIS OF URBAN/RURAL/SUBURBAN PATTERNS IN FIREFIGHTER FATALITIES

The United States Bureau of the Census defines urban as "a place having a population of at least 2,500 or lying within a designated urban area." Rural is defined as any community that is not urban. Suburban is not a census term but may be taken to refer to any place, urban or rural, that lies within a metropolitan area defined by the Census Bureau, but not within one of the central cities of that metropolitan area.

Fire department areas of responsibility do not always conform to the boundaries used for the census. For example, fire departments organized by counties or special fire protection districts may have both urban and rural coverage areas. In such cases, it may not be possible to characterize the entire coverage area of the fire department as rural or urban, and firefighter deaths were listed as urban or rural based on the particular community or location in which the fatality occurred.

The following patterns were found for 2004 firefighter fatalities. These statistics are based on answers from the fire departments and, when no data from the departments were available, the data are based upon population and area served as reported by the fire departments.

Table 20. Firefighter Deaths by Coverage Area Type

	Urban/ Suburban	Rural	Federal or State Parks/ Wildland	Total
Firefighter Deaths	58	48	11	117

IN CONCLUSION

2004 was another year of unacceptable loss for the fire service, for firefighters who lost their brothers and sisters in 2004, and—most of all—for the families and friends of the firefighters who died in 2004.

The fire service is currently engaged in an introspective look at the causes of firefighter fatalities and methods to reduce or to eliminate firefighter fatalities in the future. The National Fallen Firefighters Foundation is leading an effort to involve the entire fire service community in the solutions to the firefighter fatality problem.

The Foundation is working from a list of initiatives that was developed at a large gathering of fire service safety experts in Tampa, Florida, in the spring of 2003. The initiatives developed at that meeting were to:

1. Define and advocate the need for a cultural change within the fire service relating to safety, incorporating leadership, management, supervision, accountability, and personal responsibility.

2. Enhance the personal and organizational accountability for health and safety throughout the fire service.

3. Focus greater attention on the integration of risk management with incident management at all levels, including strategic, tactical, and planning responsibilities.

4. Empower all firefighters to stop unsafe practices.

5. Develop and implement national standards for training, qualifications, and certification (including regular recertification) that are equally applicable to all firefighters, based on the duties they are expected to perform.

6. Develop and implement national medical and physical fitness standards that are equally applicable to all firefighters, based on the duties they are expected to perform.

7. Create a national research agenda and data collection system that relate to the initiatives.

8. Use available technology wherever it can produce higher levels of health and safety.

9. Thoroughly investigate all firefighter fatalities, injuries, and near misses.

10. Ensure that grant programs support the implementation of safe practices and/ or mandate safe practices as an eligibility requirement.

United States Firefighter Disorientation Study:

A report prepared by Captain William Mora of the San Antonio Fire Department analyzed 17 incidents where firefighters became disoriented in structures, that resulted in a total of 23 firefighter fatalities.

The study revealed a disorientation sequence common to all of these incidents that included light smoke showing upon arrival, an aggressive interior attack, deteriorating interior conditions, and firefighters becoming separated from handlines.

The study recommends SOP's and training that includes a cautious initial assessment and a managed initial attack, if warranted.

The full study is available at: http://www.sanantonio.gov/safd/pdf/FirefighterDisorientationStudy.pdf or through a link at the USFA firefighter fatality Web site http://www.usfa.fema.gov/fatalities/

11. Develop and champion national standards for emergency response policies and procedures.

12. Develop and champion national protocols for response to violent incidents.

13. Provide firefighters and their families access to counseling and psychological support.

14. Provide public education with more resources and champion it as a critical fire and life safety program.

15. Strengthen advocacy for the enforcement of codes and the installation of home fire sprinklers.

16. Make safety a primary consideration in the design of apparatus and equipment.

The Foundation has devoted a great deal of resources to this effort. The process continues to develop. Additional information on this effort can be found at the Foundation's Web site at *www. firehero.org*

Something can be learned from each firefighter fatality. We owe it to the firefighters who gave their lives while onduty to seek out the lessons in their deaths and to pass that information on to other firefighters to prevent future tragedies.

The following section of this report provides information on two approaches for preventing firefighter deaths; one approach is long-term, and the other can have an immediate impact on safety. The solution to the firefighter fatality problem is not contained in any one tactical change or policy. It is going to take a concerted effort by the fire service over a period of time to make major change. Starting with the easy stuff will provide immediate benefits, but a commitment for the long term will be needed to address some causes of death.

The last firefighter to die in 2004 was Firefighter Jared Michael Moore of Kansas. His car was struck by a law enforcement vehicle while responding to a motor vehicle crash.

Opposite page: Firefighters carry the body of Firefighter Jaime Leah Foster at her funeral

SPECIAL TOPICS

EATING RIGHT

If the horrible toll of September 11, 2001, is set momentarily aside, 1,047 firefighters have died while onduty in the United States in the last decade. 460 of those firefighters died from heart attacks, the largest nature of fatal injury category for firefighter fatalities. These firefighters ranged in age from their 20's through their 80's with the largest number of them in their 50's.

In general, mankind, since the improvement of cookery, eats twice as much as nature requires.
~ Benjamin Franklin

That's a firefighter death due to a heart attack every 8 days. In addition to the deaths caused by heart attacks, the NFPA reports an average of just over 300 nonfatal heart attacks are suffered by firefighters each year while working on the scene of a fire incident in the last decade. Many of these illnesses are career-ending and life-changing, whether the firefighter is volunteer or career.

According to the American Heart Association (AHA), 130 million Americans are overweight and nearly 62 million of them are obese. People who are overweight or obese are more likely to develop heart disease and stroke, even if they have no other risk factors.

The AHA says that obesity is unhealthy because excess weight increases the strain on the heart. Obesity is linked with coronary heart disease mainly because it raises blood pressure and blood cholesterol and can make it more likely to develop diabetes.

Some risk factors for heart attack, such as advancing age and male gender, cannot be controlled. However, many heart attack risk factors such as blood pressure, blood cholesterol, smoking, and physical activity are much more within an individual's control to manage.

Healthy eating can have a positive impact on body weight, nutrition, blood pressure, and the control or avoidance of chronic diseases such as diabetes. It can be an important part of an overall plan for a healthy life that includes exercise, smoking cessation, and other risk controlling measures.

Firefighters face unique challenges when it comes to food. Meal times can be compressed by training and emergency responses. Meals are communal events that bring firefighters together, both career firefighters while onduty in their fire stations, and volunteer firefighters for fundraising and social events. Part of the fire service culture revolves around food preparation and eating together. Taking meals together is a time that firefighters have traditionally used to bond with one another and build the teamwork that is needed during response to emergency incidents.

Don't dig your grave with your own knife and fork.
~ English Proverb

Eating healthy is not a simple prospect for anyone. As Americans, we are faced with an almost unimaginable array of food choices. We are bombarded with food advertising almost from the day that we are born.

> *If food is your best friend, it's also your worst enemy.*
> ~ "Grandpa" Edward Jones, 1978

Healthy eating is not something that can be done only while onduty, it has to become a way of life for firefighters and those that they share meals with when they are offduty. This section offers some simple basics for healthy eating that will contribute to an overall healthier firefighter.

Plan at Work, Plan at Home – Convenience foods, especially prepackaged foods, tend to be less healthy than other foods. It's often easier to stop by a fast food restaurant and grab a bite on the run than go through the effort of planning and preparing a meal. Proper planning in advance for the meal can diminish some of the allure of fast food. If a crew knows in advance that they will be away from the fire station at lunch time, they may choose to prepare and pack a lunch to bring along rather than stopping for fast food on the run.

The key to healthy eating in the fire station is the cook. The cook may be a member of the onduty crew in a career fire station, a volunteer with a special flair in the kitchen, or a member of a fire department support group. If firefighters hope to have input into what is being prepared, they have to do so through the cook. It is a good idea to offer alternatives to the cook for healthier meals. The International Association of Firefighters (IAFF) recently launched a Web site called "Fit to SURVIVE." This site offers advice on fitness and recipes for more healthful meals. Use of the site is free, and site can be accessed at http://www.foodfit.com/iaff/

Planning meals is also important at home and while offduty. If the firefighter is also the cook at home or if another family member does the cooking, it is important to carry healthful eating habits started at the fire station to the home. Even for a career firefighter, most meals are eaten away from the fire station so these meals play an important role.

Read the Label – When choosing foods to eat or shopping for ingredients to be included in a recipe, read the Nutrition Facts Label. Don't be fooled by claims made on the front of the box as to the healthiness of the food, look at the label. The food label is presented in a standard format and provides information on the nutritional value of the food.

When reading food labels, pay attention to the serving size and the number of servings in the package. That little snack bag that you have chosen to eat may actually contain more than one serving of the food and that affects the amount of calories and nutrients that you will be eating.

More information about reading and interpreting food labels can be found at http://www.cfsan.fda.gov/~dms/foodlab.html

New Pyramid - In early 2005, the Food and Drug Administration (FDA) released a revised version of the Food Guide Pyramid. The FDA has provided nutritional advice for  more than 100 years, the original pyramid was developed in 1992. The new pyramid was developed to improve the effectiveness of the pyramid as a motivator for consumers to make healthier food choices and to reflect the latest nutritional science.

Along with the release of the new pyramid, the FDA launched a Web site filled with advice on healthy eating and exercise. The site includes interactive tools and the information provided by the site can be personalized to each user. The FDA site can be found at http://www.mypyramid.gov/

SIMPLE SAFETY

Many of the situations in which firefighters die are somewhat out of control – such as a structure fire. Incident Command Systems attempt to organize our response to fires and other incidents to bring some measure of control to the chaos. We use risk management planning in advance of the emergency and at the scene of an emergency to measure the risk of action against the likely benefit of those actions.

Some firefighter deaths come in extremely hard to predict situations, such as building collapses, rapid changes in fire behavior, and extreme weather. Most of the time, however, there are simple steps to safety that – if implemented – could save the lives of firefighters every year.

This section will propose some simple steps to safety during the emergency response such as firefighters who do not feel well while onduty, floor collapses, and working near moving traffic.

EMERGENCY RESPONSE

The fire service is in the emergency response business. We begin our response to every incident with the belief that our rapid arrival on the scene will contribute to a positive outcome for those impacted by the emergency situation. For this reason, we choose to accept the risks that are inherent in an emergency response.

Tasks such as driving emergency vehicles at speeds higher than the posted speed limit, negotiating through traffic, and proceeding through intersections are risky. While local laws and procedures vary, firefighters are generally allowed to move through traffic during an emergency response in ways that normal traffic is not allowed to.

Each year, firefighters are killed when their vehicles are involved in crashes. Most firefighter deaths in vehicle crashes are in personally owned vehicles. As stated earlier in this report, 58 firefighters have died in personal vehicle crashes since 1990. Each year, firefighters also die in crashes that involve fire apparatus such as pumpers, tankers, and brush trucks.

Three common threads can be seen in many of these fatal crashes; excessive speed, driving off of the right side of the roadway, and a lack of seatbelt use.

Excessive Speed – Most State laws and local fire department standard operating procedures (SOP's) allow firefighters to exceed the posted speed limit while responding to emergencies. This privilege should be managed to make sure that the risk of higher speeds provides a reasonable benefit to those needing assistance. Some suggestions for emergency response:

Limit speeds, under any circumstances, to 10 miles-per-hour above the posted speed limit. Especially in fire apparatus, speeds that are significantly above the posted speed limit are dangerous. Stopping distances are increased dramatically and high vehicle speeds in curves often have negative outcomes. This speed limit maximum should apply equally to fire apparatus, ambulances, and personal vehicles.

Wheels Off the Right – An analysis of fatal firefighter vehicle crashes reveals that one of the most common scenarios in these crashes starts with the right-hand wheels of the vehicle leaving the paved surface of the road.

The reason for the wheels leaving the roadway varies. In some situations, the side of a poorly maintained road collapses under the weight of the vehicle. Sometimes the vehicle leaves the road due to driver inattention or the fact that the driver of the vehicle is focusing his or her attention on something inside of the vehicle such as the radio or warning lights. A far more common reason is excessive speed. Especially on curves, the driver in unable to hold the vehicle on the road and the right wheels go off. Speed also contributes to situations where wheels leave the road without the complications of a curve.

Once a vehicle's right wheels leave the road, the natural inclination of most drivers is to do what they would do in their personal car or family vehicle – jog the steering wheel sharply to the left for a moment to bring the vehicle back on the road. While this likely works well with most personal vehicles, the results for a larger vehicle can be disastrous.

The center of gravity for fire apparatus is much higher than it is for passenger cars. When the driver of a piece of fire apparatus jogs the wheel sharply in any direction, the vehicle reacts much differently than a smaller vehicle. In situations where the vehicle's right wheels go off of the road and the driver jogs the steering wheel sharply to the left, the vehicle often comes back onto the road and veers into the oncoming lane of traffic. The driver again jogs the wheel sharply to the right to bring the apparatus back into its lane – many times the left side of the apparatus comes around and the vehicle begins to slide or roll. Once this critical loss of control occurs, the vehicle may crash on the road surface or leave the roadway and hit objects on either side of the road.

Wheels Off of the Road – Slow Then Steer.
When the wheels of a vehicle leave the paved surface of the roadway, it's much safer to slow down as much as possible with the wheels on the shoulder and then bring the vehicle back onto the road.

Seatbelts – The number of firefighters who are killed in otherwise survivable vehicle crashes each year is an outrage. As firefighters who respond to vehicle crashes, we know the value of seatbelts, yet we somehow cannot project those benefits onto ourselves.

In 2004, two firefighters who were riding in the open beds of pickup trucks died as the result of falls from the trucks during nonemergency circumstances. One firefighter died when she fell from the back step of an engine as it backed out of an emergency scene.

If a vehicle is not equipped with properly operating seatbelts, it should be taken out of service until the belts are provided. No firefighter should ride on a moving vehicle unless in a designated seat with a properly operating seatbelt.

No Belt – No Ride. If seatbelts are not provided, don't ride in any vehicle. If seatbelts are provided, use them.

FIREFIGHTERS FEELING UNWELL ONDUTY

In 2004, there were at least six firefighter deaths where the firefighter complained of not feeling well on the emergency scene or after returning to the fire station. These firefighters went home and then suffered fatal heart attacks within a short period of time after arriving home. In some of these cases, these firefighters were exhibiting classic signs of cardiac distress while onduty.

Many firefighters will minimize their injuries and illnesses, especially when talking with other firefighters or while onduty. If a firefighter is feeling unwell on the scene of an emergency, other firefighters should urge or compel the sick firefighter to seek immediate appropriate medical care.

THE COLLAPSE OF FLOORS

In the past decade, a number of firefighters have died when the floors of structures collapsed and propelled them into a basement that was involved in fire.

Firefighters inherently do not trust the safety of the roof of a fire-involved structure. It's common to see firefighters testing or "sounding" a roof as they conduct ventilation duties. It's far less common to see firefighters concerned with the integrity of floors in fire-involved structures.

Firefighters in Alabama, New York, North Carolina, and Ohio, to name just four examples, have died when floors in residential structures experiencing fires in their basements have failed. These fires are typically well-advanced and have been burning for some time. These collapses may be more prominent in newer structures where lightweight building materials have been used in the place of traditional full-thickness lumber.

Don't Trust the Floors – If a fire is working inside of a structure, the entire structure is simultaneously weakening – including the floors. It's best to crawl through a fire-involved structure, feeling ahead of yourself to make sure that the floor is intact. Sound floors as you would a roof to check for integrity.

WORKING NEAR TRAFFIC

Each year, firefighters are killed when they are struck by passing vehicles at the scene of an emergency. Drivers may be blinded by emergency lighting, be more concerned with getting a look at the scene than paying attention to the road, or impaired by alcohol or drugs.

Firefighters can increase their chances of survival by using their vehicles as shields at the scene, working with law enforcement officials to close roadways or lanes of the road to provide for safety, and managing onscene lighting at nighttime emergencies.

A Web site specifically developed to prevent firefighter deaths from passing traffic has been developed. The site contains sample SOP's, news, and training materials. Please see www.respondersafety.com

Act Like All Traffic is Out to Kill You – If firefighters approached their interactions with moving traffic as if all drivers passing the scene were on a mission to kill them, we would have fewer firefighter deaths each year. Firefighters need to structure the emergency scene to provide protection, work with law enforcement to make the scene safer, and then remain constantly aware of their proximity to traffic.

SUMMARY

The effort to reduce and eliminate firefighter fatalities while onduty must be accomplished with a combination of long-term and immediate steps. Improvements in the overall health of firefighters and the management of the level of risk while responding and when on the scene of an emergency can both contribute to this goal.

APPENDIX A
SUMMARY OF 2004 INCIDENTS

January 4, 2004 – 0533hrs
Leslie W. Gant, Jr., Lieutenant
Age 46, Volunteer
Winslow Township Fire Department - Sicklerville
Fire Company, New Jersey

Lieutenant Gant and members of his fire department responded to a crash on the Atlantic City Expressway. Lieutenant Gant was the driver of an engine company. Firefighters found a vehicle overturned in the woods with a report of one unaccounted-for occupant. Lieutenant Gant assisted by using a chain saw at the scene. After returning to the fire station, Lieutenant Gant experienced some dizziness and left to rest at home.

When he arrived home, Lieutenant Gant told his wife that something was wrong. His wife took him to the hospital where he was treated for anxiety and sent home. At 1138hrs, EMS was summoned to Lieutenant Gant's home. He was found to be unresponsive and was transported to a local hospital. He was transferred to a regional hospital in Philadelphia where he was diagnosed as suffering a stroke.

Lieutenant Gant lost consciousness at the hospital. Aggressive measures were taken to save his life. Despite treatment at the hospital, Lieutenant Gant died on January 8, 2004.

January 8, 2004 - 650hrs
Derrick T. Harvey, Lieutenant
Age 45, Career
Philadelphia Fire Department, Pennsylvania

Lieutenant Harvey and the members of his engine company were dispatched with other companies to the report of a structure fire in a two-story row house. Upon arrival, Lieutenant Harvey entered the structure without his self-contained breathing apparatus (SCBA) to investigate. He reported a fire in the basement to the Incident Commander (IC).

Lieutenant Harvey ordered the members of his crew to advance a hoseline into the front door of the structure. Firefighters reported a very hot and spongy floor as they advanced. Firefighters at the basement door at the rear of the structure were delayed because of accumulated debris.

Firefighters operating the handline from Lieutenant Harvey's engine were directed by the IC to prevent the fire from extending into the first floor from the basement. Lieutenant Harvey was working inside with his firefighters, still without the benefit of a SCBA. As Lieutenant Harvey exited the structure for a third time, he fell partially through the floor and was trapped.

Firefighters from Lieutenant Harvey's crew became concerned when they did not hear from him. One firefighter crawled back toward the front of the home but could not reach the front due to floor failure. Firefighters were able to exit the building through the basement; fire control was being achieved.

Once outside, firefighters informed the IC that Lieutenant Harvey was missing. A rapid intervention crew entered the structure and located Lieutenant Harvey. He was found face down with one leg through a hole in the floor. At this point, approximately 18 minutes had passed since Lieutenant Harvey left his crew.

continued on next page

43

Lieutenant Harvey was removed from the building and transported to the hospital. Lieutenant Harvey suffered from burns on approximately 32 percent of his body, smoke inhalation, and carbon monoxide inhalation. He was treated at the hospital and then transferred to a burn unit. Despite aggressive treatment for his injuries, Lieutenant Harvey died on January 15, 2004. The cause of death was listed as smoke and soot inhalation and burns.

The fire was caused when combustibles were stored too close to an oil-burning heater.

For additional information regarding this incident, please refer to NIOSH Firefighter Fatality Investigation and Prevention Program report F2004-05 (www.cdc.gov/niosh/face200405.html).

January 9, 2004 – 1043hrs
Gregory Harold Vieth, Lieutenant
Age 53, Career
Davenport Fire Department, Iowa

While on duty, Lieutenant Vieth complained to other firefighters that he was not feeling well. After going off duty, he was completing regular physical fitness exercises when he experienced an unwitnessed collapse at approximately 1043hrs.

Passersby noticed Lieutenant Vieth lying on the sidewalk near his home and called 9-1-1. Responding firefighters found Lieutenant Vieth pulseless and not breathing. Advanced life support (ALS)-level care was provided, and Lieutenant Vieth was transported to the hospital. Lieutenant Vieth did not respond to any treatments and was pronounced dead as a result of a heart attack at 1211hrs.

January 15, 2004 – 0500hrs
Richard Allen Jones, Firefighter
Age 65, Volunteer
Maryland Line Volunteer Protection Association - Baltimore County Fire Dept. Maryland

Firefighter Jones responded to two fire incidents on January 14, 2004. For each incident, Firefighter Jones responded to the fire station to staff equipment on a standby basis. He returned home and went to bed at the end of the day. Firefighter Jones died during the night of hypertensive atherosclerotic cardiovascular disease (a heart attack).

January 21, 2004 – 2111hrs
Keith Allen Firment, Captain
Age 39, Volunteer
Marguerite Volunteer Fire Department, Pennsylvania

Captain Firment had been working on restoring heat to a portion of the fire station. This task involved digging several feet into the ground to reach an oil tank.

Captain Firment responded to a motor vehicle crash and assisted with extrication at the scene.

Captain Firment and the members of his fire department later responded to reports of a structure fire. Once on scene, the incident was found to be a false alarm. Upon returning to the fire station, Captain Firment told another firefighter that he was not feeling well and went home.

Within a few minutes of his arrival at home, Captain Firment collapsed. He was transported to the hospital and died the next day. The cause of death was listed as a ruptured aorta.

January 22, 2004 – 1214hrs
Charles "Charlie" T. Hatch, Jr., Firefighter/Paramedic
Age 48, Career
West Bridgewater Fire Department, Massachusetts

On January 21, 2004, at 0958hrs, Firefighter Hatch and other offduty members of his department were called in for fire station coverage as onduty members of his department responded to a mutual-aid structure fire.

While on standby in the fire station, Firefighter Hatch stocked a new ambulance and performed other physical tasks involving lifting and moving heavy objects. When onduty firefighters returned to quarters, Firefighter Hatch was released from duty. The time of the release was approximately 1428hrs.

On January 22, 2004, at approximately 1214hrs, Firefighter Hatch was working in the shop of his personal business. He complained of not feeling well and collapsed. A bystander performed cardiopulmonary resuscitation (CPR) until paramedics arrived. Firefighter Hatch did not recover and died later that day as the result of a heart attack.

Firefighter Hatch became ill approximately 21 hours after being released from standby duty. Firefighter Hatch had a history of heart difficulties but was certified for duty by a physician in December of 2003.

For additional information regarding this incident, please refer to NIOSH Firefighter Fatality Investigation and Prevention Program report F2004-08 (www.cdc.gov/niosh/face200408.html).

January 24, 2004 – 1544hrs
Kevin Michael Shea, Fire Chief
Age 54, Volunteer
Elsmere Fire Department, New York

Chief Shea and members of his fire department responded to a structure fire alarm in a local nursing home. The alarm was unfounded and all units returned to the fire station.

As firefighters were leaving the station, they noticed that Chief Shea was leaning over the center console in his vehicle. Firefighters discovered that Chief Shea was unconscious and in distress. Chief Shea was removed from the vehicle, and CPR was started. Upon arrival of emergency medical services (EMS) and paramedics, advanced cardiac life support (ACLS) measures were initiated, and Chief Shea was transported to the hospital.

Despite treatment in the ambulance and at the hospital, Chief Shea was pronounced dead. The cause of death was listed as a heart attack.

January 27, 2004 – 1432hrs
David Andrew Mackie, Firefighter/Paramedic
Age 35, Career
Orange City Fire Department, Florida

Firefighter/Paramedic Mackie was attending a class at a local community college that would lead to State firefighter certification. He had been attending training all day and participated in a light jog after lunch.

Firefighter/Paramedic Mackie returned to the classroom and was there for a short period of time when he collapsed. Other students and staff provided emergency medical care, including CPR and the use of a defibrillator. Firefighter/Paramedic Mackie was transported to the hospital where efforts to revive him were continued for 30 minutes. These efforts were unsuccessful, and Firefighter/Paramedic Mackie was pronounced dead.

An autopsy revealed that Firefighter/Paramedic Mackie had a problem with a heart valve that led to his collapse and death. Cardiac abnormalities had been noted in previous physical examinations, but Firefighter/Paramedic Mackie had been cleared for duty.

For additional information regarding this incident, please refer to NIOSH Firefighter Fatality Investigation and Prevention Program report F2004-12 (www.cdc.gov/niosh/face200412.html).

January 28, 2004 – 0700hrs
Jean L. Nuckols, Firefighter
Age 47, Career
Navy Regional Fire Rescue, Norfolk, Virginia

Firefighter Nuckols suffered a stroke (CVA) while on duty in the fire station. He had participated in strenuous training activities while on duty the day before his collapse.

Firefighter Nuckols was transported to the hospital and died on January 31, 2004.

~~~

February 3, 2004 – 1754hrs
Michael Edward Lynch, Firefighter
Age 32, Volunteer
Penrose Volunteer Fire Department, Colorado

Firefighter Lynch was directing traffic at the scene of an earlier motor vehicle crash on a median-divided highway with two lanes in each direction. He was equipped with a traffic vest, a stop/slow sign, and a flashlight.

Firefighter Lynch was in the right-hand lane near a vehicle that had slowed abruptly in the left lane. A pickup truck approached the scene in the left lane, and then swerved into the right lane to avoid hitting the vehicle in the left lane. The pickup truck struck Firefighter Lynch and threw him off of the right side of the road. He came to rest 136 feet from the point of impact.

Firefighter Lynch was transported to the hospital and was pronounced dead. The cause of death was listed as major trauma.

The driver of the pickup truck, a minor, pleaded guilty to a charge of careless driving involving a death.

~~~

February 4, 2004 – 1500hrs
Glenn Christopher "Galdo" Galderisi, Firefighter
Age 52, Volunteer
Wayne Township, Pompton Falls Volunteer Fire Department #3, New Jersey

Firefighter Galderisi and members of his department had returned to the station after a response to a false alarm. Firefighter Galderisi adjusted some hose on the apparatus that had shifted during the response and then suffered a heart attack.

Firefighters immediately provided assistance, including CPR, and summoned an ambulance. Firefighter Galderisi was transported to a local hospital where he was pronounced dead.

47

February 13, 2004 – 1545hrs
Brenda Denise Cowan, Lieutenant
Age 40, Career
Lexington Division of Fire & Emergency Services, Kentucky

Lieutenant Cowan and her engine company were dispatched with an ambulance to a report of a possible shooting in which a female subject was down and bleeding from the head. A neighbor had reported the incident.

The engine was staged away from the scene, and Lieutenant Cowan and a member of her crew walked up the road to attempt to locate the victim. Lieutenant Cowan and the firefighter located the victim and assessed injuries. The victim appeared to be expired, and an automatic external defibrillator (AED) was being applied to confirm death. The ambulance arrived at the scene and drove to the victim's location.

As the ambulance arrived, shots were fired at the responders. Lieutenant Cowan and the firefighter were struck and took cover behind a tree. Lieutenant Cowan was struck a second time.

A police officer who had responded to the initial call attempted to provide cover for fire and EMS responders by placing his vehicle between the source of the gunfire and the responders. His vehicle also took fire.

Once sufficient law enforcement personnel had arrived on the scene, the firefighter and Lieutenant Cowan were extricated from the scene by a police rescue team. Once out of the danger area, life-saving procedures were initiated on Lieutenant Cowan. She was transported by ambulance to a local hospital where she was pronounced dead from her wounds.

The firefighter's injuries were less severe, and he was released from the hospital 2 days after the incident.

The initial shooting was related to a domestic dispute. The husband of the initial victim was taken into police custody approximately 6 hours after the start of the incident. He was charged with 2 counts of murder and 2 counts of attempted murder. Prosecution of the charges has been delayed due to the mental condition of the alleged assailant.

February 14, 2004 – 0715hrs
Robert E. Heminger, Captain
Age 39, Volunteer
Wood River Fire & Rescue, Nebraska

Kenneth A. Woitalewicz, Captain
Age 38, Volunteer
Wood River Fire & Rescue, Nebraska

Captain Heminger and Captain Woitalewicz responded with other members of their fire department to a structure fire in a single-story residence. Upon their arrival on the scene, firefighters discovered a working fire and received reports of a trapped occupant.

Captain Heminger and Captain Woitalewicz advanced an attack line into the residence for search and rescue and fire control. During the search, the firefighters entered a room that had been added to the home. Without warning, the roof of the addition collapsed on the firefighters in a pancake-type collapse. Both firefighters were trapped in the collapse that occurred 17 minutes after they arrived on scene.

Firefighters were unable to see either Captain under the collapsed roof, but personal accountability safety system (PASS) device alarms could be heard. A front-end loader was brought to the scene and was used to lift the roof far enough for access to the trapped firefighters. Both firefighters were removed and found to be in respiratory and cardiac arrest. Both were transported to a local hospital. The firefighters had been trapped for approximately 12 minutes.

Captain Heminger died the following morning, and Captain Woitalewicz died the afternoon of February 17, 2004. Both deaths were attributed to asphyxia/anoxia secondary to their entrapment.

The collapse was caused by multiple factors including a buildup of ice on the roof of the addition, an unsloped roof on the addition, rusted fasteners used to attach the addition to the original structure, poor construction practices, and fire exposure. The occupant who was trapped in the structure died of smoke inhalation. The deceased occupant had been on oxygen for a medical condition, and the presence of supplemental oxygen supplies in the home was thought to have contributed to the intensity of the fire.

February 16, 2004 – 1117hrs
Ernest "Ernie" Heatherman, Fire Police Captain
Age 45, Volunteer
Brisbin Fire Department, New York

Fire Police Captain Heatherman and members of his fire department responded to a mutual-aid chimney fire. Upon their return to the fire station, Fire Police Captain Heatherman complained of not feeling well and told the fire chief that he was going home.

Approximately 45 minutes after leaving the fire station, Fire Police Captain Heatherman suffered a heart attack. Firefighters and emergency workers responded to his residence and provided aid. Fire Police Captain Heatherman was transported to the hospital where he was later pronounced dead.

February 18, 2004 – 1345hrs
Steven Wayne Fierro, Firefighter
Age 40, Career
Carthage Fire Department, Missouri

Firefighter Fierro and members of his engine company were dispatched to a mutual-aid structure fire involving a large bar. Once on scene, firefighters observed heavy smoke but could not locate the source of the fire.

Firefighter Fierro was equipped with an SCBA, integrated PASS device, and a helmet-mounted thermal imaging camera. He entered the structure with two other firefighters on an attack line to search for the seat of the fire. Initially, conditions were clear with only light smoke and good visibility. Conditions worsened rapidly, and visibility in the structure was reduced to zero.

After one of the firefighters with Firefighter Fierro experienced a problem with his SCBA, he was forced to exit the building. The other firefighter with Firefighter Fierro also exited, believing that he had been left alone.

Firefighters and the IC on the exterior of the structure noted structural failure and ordered all firefighters out of the building. Air horns were sounded to signal the evacuation, and the handline that had been advanced by Firefighter Fierro was pulled from the building.

Firefighters noticed that Firefighter Fierro was missing and initiated searches of the building. Approximately 56 minutes after his initial entry, Firefighter Fierro was discovered approximately 25 feet from the entrance. He was found face-down and had been severely burned. Firefighter Fierro's SCBA and wires from the thermal imaging camera were found entangled in a chair. According to reports, the entanglement was likely not significant enough to prevent Firefighter Fierro's escape. Firefighter Fierro was obviously dead and was later removed from the structure.

The cause of death for Firefighter Fierro was listed as smoke inhalation. His blood carbon dioxide level was found to be 51 percent.

For additional information regarding this incident, please refer to NIOSH Firefighter Fatality Investigation and Prevention Program report F2004-10 (www.cdc.gov/niosh/face200410.html).

February 22, 2004 – 1430hrs
Elliott Davis, Jr., Fire Commissioner
Age 51, Volunteer
Gloster Rural Volunteer Fire Department, Mississippi

Fire Commissioner Davis was driving a 1997 GMC C-8500 water tanker headed to a grass fire. As the apparatus entered a left-hand curve at 50 to 55 miles per hour, Fire Commissioner Davis moved to the right to avoid hitting a car that was approaching from the opposite direction.

When the right wheels of the apparatus left the right side of the roadway, Fire Commissioner Davis overcorrected to the left causing the left wheels of the apparatus to leave the left side of the road. Fire Commissioner Davis corrected again, and the apparatus began to roll.

Fire Commissioner Davis was partially ejected from the apparatus and was pronounced dead at the scene. A passenger in the truck was treated for non-life-threatening injuries. Neither firefighter was wearing a seatbelt at the time of the crash. The passenger reported to police that the apparatus had been traveling too fast.

Fire Commissioner Davis was an alderman in Gloster Ward and was in charge of the town's fire and ambulance departments.

~∞∞~

February 23, 2004 – 1330hrs
Richard L. Gabrielli, Fire Police Officer
Age 70, Volunteer
George G. McMurtry Volunteer Fire Department, Vandergrift, Pennsylvania

Fire Police Officer Gabrielli was directing traffic at the entrance to a steel plant. A working fire was in progress within the plant site, and Fire Police Officer Gabrielli was present to direct responding fire apparatus into the scene.

Witnesses observed Fire Police Officer Gabrielli collapse onto the street. An ambulance that routinely responds on fire calls was summoned, and Fire Police Officer Gabrielli was transported to the hospital.

Fire Police Officer Gabrielli died as the result of a heart attack.

~∞∞~

February 23, 2004 – 1630hrs
Edward P. Conricote, Firefighter
Age 55, Volunteer
Liberty Fire Department, Ohio

Firefighter Conricote responded to the scene of a residential structure fire in his personal vehicle and arrived just before the first engine company. Firefighter Conricote assisted the engine's driver/operator with supply line hookups to a fire hydrant and opened the hydrant.

A police officer observed Firefighter Conricote trip over an object and asked if he was alright. Firefighter Conricote replied that he was fine and continued working. A firefighter left the fire-involved structure in search of a pike pole to hold open an overhead garage door. Firefighter Conricote stepped up onto the back of the engine to retrieve the tool.

Firefighter Conricote suddenly bent over and wrapped his arms around his midsection. The other firefighter asked if he was alright and received no answer. Firefighter Conricote fell backwards and landed on the roadway.

continued on next page

51

Firefighter Conricote was treated at the scene by firefighters and EMS workers and then transported to a local hospital. He did not respond to treatment and was pronounced dead at the hospital. The cause of death was listed as a heart attack.

For additional information regarding this incident, please refer to NIOSH Firefighter Fatality Investigation and Prevention Program report F2004-46 (www.cdc.gov/niosh/face200446.html).

~~~

February 23, 2004 – 1815hrs
Bret Eugene Neff, Deputy Chief
Age 37, Volunteer
Harford Fire Company, New York

Deputy Chief Neff was the passenger in a water tanker that responded to a mutual-aid structure fire. Upon their arrival on the scene of the incident, Deputy Chief Neff's tanker was directed to fill a portable water tank that was supplying water for the fire attack.

The driver of the tanker backed the truck into position for a rear water dump. Deputy Chief Neff went to the rear of the vehicle to open the valve. As the driver exited the vehicle to join Deputy Chief Neff, the driver accidentally released the emergency brake. The truck rolled backwards and crushed Deputy Chief Neff between the tanker and the apparatus that was drafting from the portable tank.

The driver returned to the cab of the tanker, placed the apparatus in gear, and rolled forward slightly before rolling backward again and pinning Deputy Chief Neff again. The tanker was moved forward and firefighters began to assess the injuries suffered by Deputy Chief Neff. EMS personnel treated him at the scene.

Deputy Chief Neff was then transported by helicopter to a hospital where he was aggressively treated for his injuries. Despite efforts at the scene, in the helicopter, and in the emergency room, Deputy Chief Neff died at the hospital.

The cause of death was listed as blunt force trauma to the abdomen.

~~~

March 3, 2004 – 0912hrs
Edward "Eddie" Orlando Peters, Forest Ranger
Age 40, Wildland Full-Time
Florida Division of Forestry, Florida

Forest Ranger Peters was driving to a controlled burn from his work station in a Division of Forestry Brush Patrol Truck. The vehicle was a 1986 Chevrolet pickup truck chassis fitted with a small water tank and pump at the back of the truck.

As he drove on a State road, Forest Ranger Peters' vehicle experienced a blowout or tire separation on the right front tire. The vehicle traveled off of the right-hand side of the road, hit a culvert, and flipped upside down. Forest Ranger Peters was trapped underwater inside the vehicle and drowned. The depth of the water was approximately 2 feet; Forest Ranger Peters was wearing his seatbelt.

Other drivers who either witnessed or came upon the crash scene removed Forest Ranger Peters from the vehicle. It was estimated that he had been submerged for approximately 10 minutes. EMS responders pronounced Peters dead at the scene.

March 11, 2004 – 1445hrs
Mark Eugene Miller, Lieutenant
Age 43, Career
Laconia Fire Department, New Hampshire

Lieutenant Miller and another firefighter were at a local lake testing some new diving equipment that had been purchased by the Laconia Fire Department. Lieutenant Miller had worn the suit at least eight times before.

Lieutenant Miller was not tethered to the other diver as they began their dive in open water. The divers did not intend to make any part of the dive under ice. The divers gave each other a "thumbs up" to surface. The firefighter training with Lieutenant Miller surfaced, but Lieutenant Miller failed to surface.

A search was begun immediately but had to be stopped on account of darkness after approximately 4 hours. The search began again in the morning, and Lieutenant Miller's body was recovered at about 1300hrs.

A New Hampshire Marine Patrol investigation did not cite any procedural errors, and it did not find that any health condition or substance caused the accident.

Lieutenant Miller's dream was to establish a dive rescue team for his department. After his death, the fire chief pledged to make the team a reality. The team went into operation by the fall of 2004 thanks to the efforts of many and the support of the community.

March 13, 2004 – 0732hrs
Donald Eugene Ward, Firefighter
Age 46, Career
Columbus Division of Fire, Ohio

Firefighter Ward reported for duty for a 12-hour overtime shift on March 12, 2004. He was assigned to a heavy rescue company and ran seven emergency calls during the shift, including a working structure fire.

continued on next page

Prior to going off duty at 0800hrs on March 13, 2004, Firefighter Ward told other firefighters that he was not feeling well. During the rest of that day, Firefighter Ward told several others that he was not feeling well.

On March 14, 2004, at approximately 0830hrs, Firefighter Ward's wife discovered him at the construction site of their new home. Firefighter Ward was gravely ill and was transported to the hospital. He was pronounced dead at 0937hrs.

The cause of death was listed as hypertensive and arteriosclerotic cardiovascular disease.

March 13, 2004 – 0846hrs
Charles G. Brace, Battalion Chief
Age 55, Career
Pittsburgh Fire Bureau, Pennsylvania

Richard A. Stefanakis, Master Firefighter
Age 51, Career
Pittsburgh Fire Bureau, Pennsylvania

Members of the Pittsburgh Fire Bureau were engaged in a long-term firefight in a historic church. The fire had been underway for hours and had reached five alarms.

The fire was pretty much under control with most flame knocked down at around 1200hrs. Battalion Chief Brace and Firefighter Stefanakis were in the vestibule at the front of the church setting up lines to overhaul the fire. The basement was filled with burning debris as the result of collapses.

Without warning, at approximately 1219hrs, the church bell tower collapsed on top of the firefighters in the vestibule. Battalion Chief Brace was covered with debris in the vestibule. Firefighter Stefanakis was pushed into the basement and buried. It took approximately 2 hours to reach Firefighter Stefanakis and 3 hours to remove Battalion Chief Brace from the debris.

The cause of death for both firefighters was asphyxiation as the result of being crushed. In addition to the 2 firefighter deaths, 29 firefighters were injured.

March 13, 2004 – 1400hrs
Robert A. Griffin, Firefighter
Age 38, Volunteer
Volunteer Fire Department of Prospect, Inc., Connecticut

Firefighter Griffin was a student in a Firefighter I class at a regional training facility. Firefighter Griffin and the members of his class spent the morning donning SCBA's and doing walks in the training tower using their SCBA's. The class broke for lunch after the morning's training.

After lunch, Firefighter Griffin and the other members of his group completed a wood chopping exercise while using SCBA's and advanced to the maze station. Firefighter Griffin completed one trip through the maze and then took a break. As Firefighter Griffin neared the end of his second trip through the maze, instructors observed him having difficulty. When asked, he requested assistance. Another student and instructors assisted Firefighter Griffin and then removed him from the maze when he became unconscious.

CPR was initiated, and EMS was called to respond. Upon arrival of the ambulance, Firefighter Griffin was transported to the hospital. He was pronounced dead at the hospital after intervention efforts were attempted.

The cause of death was listed as sudden death associated with rheumatic heart disease.

March 13, 2004 – 1654hrs
Mario Felipe Cunha, Fire Engineer
Age 32, Paid-on-Call
City of Soledad Fire Department, California

Fire Engineer Cunha and the members of his engine company were preparing to fight a vehicle and brush fire on a local highway. Fire Engineer Cunha was stretching an attack line.

As Firefighter Cunha worked, a passenger car drove through the smoke from the fires and attempted to pass through the scene. The car struck Fire Engineer Cunha and threw him onto the top of the engine. Firefighter Cunha was pronounced dead at the scene.

The driver of the car was arrested and charged with several counts, including leaving the scene of the crash.

March 16, 2004 – 0935hrs
James Edward Towell, Jr., Pilot
Age 63, Wildland Contractor
Western Pilot Service under contract to the Bureau of Land Management, Arizona

Pilot Towell was participating in the Bureau of Land Management's Single-Engine Air Tanker (SEAT) Pilot Academy near Safford, Arizona. Pilot Towell was practicing a water drop under the supervision of an IC on the ground.

He requested and received permission to do a dry run prior to the water drop. As the aircraft proceeded through the dry run, it crashed. Initial unconfirmed reports blamed engine trouble. The plane was destroyed on impact, and Pilot Towell was killed.

continued on next page

For additional information about this crash, consult the National Transportation Safety Board Web site at http://www.ntsb.gov/ntsb/query.asp - NTSB identification LAX04TA161.

∿∿∿

March 16, 2004 – 2230hrs
Barrie J. Niebergall, Driver/Operator/Engineer
Age 56, Career
Red, White, and Blue Fire Protection District, Colorado

Driver/Operator/Engineer Niebergall was alone in his fire station's exercise room performing physical fitness activities. Another firefighter discovered him unconscious and not breathing. ALS-level emergency care was provided by firefighters, and additional assistance was dispatched.

No response to extensive medical treatment was realized, and permission to discontinue resuscitative measures was secured from medical control.

Driver/Operator/Engineer Niebergall's death was caused by atherosclerotic cardiovascular disease.

∿∿∿

March 17, 2004 – 1530hrs
Victor Scott, Deputy Fire Chief
Age 61, Volunteer
Otter Creek Volunteer Fire Department, Florida

Deputy Chief Scott was driving a fire department-owned pickup truck carrying fire equipment that had been donated by a nearby career fire department. Deputy Chief Scott had complained of not feeling well the night before at a fire department drill.

As he drove, Deputy Chief Scott suffered a heart attack. Deputy Chief Scott was pronounced dead at the scene.

∿∿∿

March 20, 2004 – 1019hrs
Charles Michael Lehnen, Assistant Chief
Age 57, Volunteer
Bethalto Volunteer Fire Department, Illinois

Assistant Chief Lehnen and other members of his fire department responded to a structure fire on March 19, 2004.

The next morning, Chief Lehnen awoke and complained of weakness and numbness. He was treated at a local hospital where he experienced lightheadedness and dizziness. After being awake for 2 hours, Assistant Chief Lenhen became unconscious. He was diagnosed as suffering from a CVA.

56

Assistant Chief Lehnen was transferred to a regional hospital where his condition continued to worsen. He died as a result of the CVA on March 21, 2004.

March 21, 2004 – 1421hrs
Robert Lee Smith, Firefighter
Age 68, Volunteer
West Shelby Fire and Rescue District, Alabama

Firefighter Smith and the members of his fire department responded to the scene of an intentionally set wildland fire. The fire was caused when a small brush and trash pile fire was ignited while the owners of the property were away.

The fire was brought under control. Approximately 23 minutes into the incident, Firefighter Smith collapsed of an apparent heart attack. Firefighter Smith was attended to by emergency medical technicians (EMT's) and paramedics and transported to the hospital. He was pronounced dead at the hospital.

March 21, 2004 – 1815hrs
Terri Lynn Eiland, Lieutenant
Age 40, Volunteer
Forts Lake/Franklin Creek Volunteer Fire Department, Mississippi

Lieutenant Eiland was the driver of a medium-sized brush/rescue truck (F350) responding to the report of a brush fire. An engine company was also responding to the incident but was delayed by a train.

The right front tire of the brush/rescue vehicle left the paved surface of the road, Lieutenant Eiland overcorrected to the left, and the vehicle entered into a clockwise spin. The brush/rescue truck entered a ditch and flipped end-over-end and came to rest on its roof. Another firefighter passenger was able to get out of the vehicle.

Other firefighters arrived on the scene but were not able to remove Lieutenant Eiland from the vehicle. The department's extrication tools were not accessible due to the level of damage that the brush/rescue vehicle had incurred. Mutual aid was requested for extrication.

Lieutenant Eiland was in and out of consciousness during the rescue effort. She was removed from the vehicle approximately 30 minutes after the crash and transported by ambulance to a local hospital. She was pronounced dead at the hospital. The cause of death was listed as cardiac and pulmonary results of compression asphyxia.

Lieutenant Eiland and the firefighter passenger were wearing seatbelts, and the rescue airbags operated. Excessive speed was cited in the law enforcement crash investigation.

March 25, 2004 – 0755hrs
Kenneth Michael Temke, Firefighter
Age 45. Volunteer
Alexandria Fire/EMS, Fire District #5, Kentucky

Firefighter Temke and members of his department were dispatched to a report of smoke and flames visible at a location in their jurisdiction. A fire officer near the scene responded in his personal vehicle, and two engines responded. Firefighter Temke was a passenger on the second engine.

Upon arrival on the scene, the fire officer reported a vehicle fire and released the second engine. Firefighter Temke's engine had not yet departed from the fire station; it was loaded and preparing to depart. When the response of the engine was cancelled, the engine was backed into quarters.

As firefighters dismounted the engine, Firefighter Temke fell to the ground from the jumpseat. Other firefighters went to his aid and found him to be unresponsive. An AED was applied and one shock was delivered. CPR was initiated.

Firefighter Temke was transported by basic life support (BLS) ambulance. The ambulance was met en route by paramedics, and ACLS care was provided. Despite care provided by firefighters, EMS workers, and hospital staff, Firefighter Temke was pronounced dead shortly after his arrival at the hospital.

Firefighter Temke had a history of cardiac problems. His death was caused by hypertensive and atherosclerotic cardiovascular disease.

March 26, 2004 – 2114hrs
Joseph S. Pepe III, Firefighter
Age 48, Career
Springfield Fire Department, New Jersey

Firefighter Pepe responded to an odor of natural gas incident. When the incident was concluded, Firefighter Pepe participated in a charity basketball game involving law enforcement, EMS, and fire personnel.

During the game, Firefighter Pepe suffered a seizure-like episode in the presence of EMS workers. Treatment was initiated immediately, and an AED was applied. Firefighter Pepe received two shocks from the AED, and CPR was provided. He was transported to the hospital where he died as the result of a heart attack.

58

March 28, 2004 – 1630hrs
James Harold Pennington, Fire Chief
Age 63, Volunteer
Unity-Frost Prairie Volunteer Fire Department, Arkansas

Chief Pennington and members of his department responded to a wildland fire. Chief Pennington arrived in his personal vehicle and operated the pumper when it arrived on scene. After the fire was extinguished, Chief Pennington complained of chest pains. He left the scene and went home to rest.

About an hour and a half after leaving the scene, Chief Pennington collapsed at home of an apparent heart attack. Firefighters who were at Chief Pennington's home responded immediately, and CPR was initiated. An ambulance transported Chief Pennington to the hospital where he was pronounced dead.

Children playing with matches caused the fire.

March 28, 2004 – 1900hrs
Rickford L. Wilbur, Fire Chief
Age 56, Volunteer
Eaton Rapids Township Fire Department, Michigan

Chief Wilbur and the members of his fire department responded to a report of a structure fire. When firefighters arrived on the scene, they found a controlled burn in a brush pile; the land owners were not aware that they needed a permit to burn brush.

The fire was not extinguished by the fire department and members left the scene to return to their station. Shortly thereafter, firefighters were dispatched back to the same address for a medical emergency.

Chief Wilbur had begun to leave the scene and suffered a heart attack. Firefighters had to break into his vehicle to access him. CPR was initiated and an AED was applied. He was transported by ambulance to a local hospital where he was pronounced dead.

April 1, 2004 – 1211hrs
Joseph Scott Northup, Firefighter
Age 49, Volunteer
Jessamine County Fire District, Kentucky

At 0830hrs, Firefighter Northup and members of his fire department responded to a fire call. Firefighter Northup responded in his personal vehicle, but the call was cancelled by the IC prior to reaching the scene.

Firefighter Northup was not feeling well and called his wife to say that he was coming home. Firefighter Northup went to an EMS office complaining of chest pain at approximately 1000hrs. He was placed on a cardiac monitor, and the results were normal. He was advised by EMS

to go to the hospital but refused and said that he would go in if he felt worse later.

Just after noon, Firefighter Northup's wife found him unresponsive on the couch at home. EMS was called and found Firefighter Northup pulseless and not breathing. He was treated at the scene, in the ambulance, and upon arrival at the hospital, but he could not be revived.

The cause of death was listed as a heart attack.

April 4, 2004 – 0555hrs
Kevin Wayne Kulow, Probationary Firefighter
Age 32, Career
Houston Fire Department, Texas

Firefighter Kulow and the members of his engine company responded to a structure fire in a nightclub. Upon arrival, Firefighter Kulow and other firefighters advanced a hoseline into the interior of the structure, based on reports of people trapped.

At some point in the firefight, firefighters were ordered to evacuate the building. Firefighter Kulow did not emerge and was later found inside the structure. His death was caused by burns.

The fire was caused when one or more arsonists poured and ignited accelerants in an attempt to cause the death of a nightclub employee as a part of a domestic dispute. Three individuals were charged with murder.

Firefighter Kulow had arrived early for duty and responded with the onduty crew. In addition to the death of Firefighter Kulow, three firefighters were injured while working at the scene.

Details related to this incident are not available due to pending litigation.

A report on this incident will be prepared by the Texas State Fire Marshal. The report will be available at http://www.tdi.state.tx.us/fire/fmloddinvesti.html

April 6, 2004 – 1856hrs
Phillip Stephen Hulen, Firefighter
Age 19, Volunteer
Vann Crossroads Fire Department, Inc., North Carolina

Firefighter Hulen stopped by his fire station to pick up his protective clothing while en route to firefighter training at a local community college. Firefighter Hulen was preparing to attend his last night of class for Firefighter I & II certification.

As Firefighter Hulen entered a curve in his personal vehicle, the right wheels of the vehicle left the roadway and struck a sign. The vehicle came back onto the roadway and then left the roadway again on the left side, it overturned, and then struck a tree. Firefighter Hulen was pronounced dead at the scene.

Firefighter Hulen was not wearing a seatbelt; however, an air bag was deployed as a result of the crash. The law enforcement crash report stated that the speed of the vehicle prior to the crash was 65 miles per hour in a 55 miles-per-hour zone.

~~~

April 8, 2004 – 1044hrs
Edward Hayes Stallings, Fire Chief
Age 71, Volunteer
Carthage Fire Department, Tennessee

Chief Stallings and the members of his fire department responded to a report of a structure fire in a church. A heat gun being used to remove old paint from a window had ignited materials inside the wall.

When firefighters arrived, flames were visible from the exterior of the structure. The front entry to the church was forced open, and firefighters found flames showing from an interior wall. In the meantime, the fire had extended into the attic of the church. The attic had an accumulation of approximately two inches of coal dust from the time when the church was heated with a coal furnace.

Chief Stallings was standing outside the building observing operations. Chief Stallings saw a rapidly developing fire and moved toward the building. As he was doing so, the front façade of the building fell outward and collapsed on top of Chief Stallings and two other firefighters.

Chief Stallings took the brunt of the collapse since he was further away from the building. Chief Stallings was transported by helicopter to a regional hospital. He died as a result of his injuries on August 1, 2004.

~~~

April 8, 2004 – 1122hrs
Leslie Keith Gillum, Firefighter
Age 71, Volunteer
Norton Branch Volunteer Fire Department, Kentucky

Firefighter Gillum and the members of his department responded to a motor vehicle crash to assist with extrication. Firefighter Gillum set up the department's hydraulic extrication tool and then began to experience severe chest pains.

Firefighter Gillum was evaluated by emergency medical personnel on the scene and advised that he should be transported by ambulance to the hospital. Firefighter Gillum refused transportation and assisted with

continued on next page

the set up of a landing zone for a helicopter. After the landing zone was set, Firefighter Gillum signed a refusal for treatment with EMS.

At the conclusion of the incident, Firefighter Gillum was driven to the hospital by his wife. Firefighter Gillum was admitted to the hospital and was scheduled for heart surgery. He died during the surgery on April 14, 2004.

~~~

April 10, 2004 – 1300hrs
Kenneth Eugene Sterling, Firefighter/EMT
Age 43, Career
Westview-Fairforest Fire and EMS Department, South Carolina

Firefighter/EMT Sterling responded to an incident at approximately 0700hrs. After returning from the incident, Firefighter Sterling went off duty.

At approximately 1300hrs, Firefighter Sterling was driving his personal vehicle when the vehicle swerved off of the road and came to rest. Firefighter Sterling was observed slumped over the steering wheel. Local firefighters responded but were unable to revive Firefighter Sterling. His death was caused by a heart attack.

~~~

April 15, 2004 – 0930hrs
Michael Joseph Fenster, Acting Fire Chief
Age 57, Career
Capital City Fire & Rescue, Alaska

Chief Fenster became ill at home shortly after leaving work. He called the fire station and requested that firefighters respond to his residence to treat him since he did not feel that he could drive.

Firefighters arrived and found Chief Fenster in cardiac distress. He was treated with ALS protocols and transported to the hospital. Upon arrival at the hospital, Chief Fenster suffered a heart attack. Despite efforts to revive him, Chief Fenster was pronounced dead.

For additional information regarding this incident, please refer to NIOSH Firefighter Fatality Investigation and Prevention Program report F2004-24 (www.cdc.gov/niosh/face200424.html).

~~~

April 18, 2004 – Time Unknown
Kevin R. McIntyre, District Chief
Age 45, Career
Rockford Fire Department, Illinois

District Chief McIntyre reported for duty on April 18, 2004. During the day, he served as the IC for a dive-rescue drill at a local dam. About 1800hrs, Chief McIntyre called the fire department alarm office and informed them that he was not feeling well and that he was going home. District Chief McIntyre told others that he was experiencing back pain.

An onduty captain drove District Chief McIntyre to his home. At approximately 1700hrs on April 19, 2004, a friend found District Chief McIntyre deceased in his home. The cause of death was a heart attack.

April 22, 2004 – 1558hrs
Edgar Bruce Rogers, Lieutenant
Age 56, Volunteer
Chesterfield Fire Department, South Carolina

Lieutenant Rogers and members of his fire department were fighting a major fire in a community center. Lieutenant Rogers was assisting with hoseline deployment on a sloping hillside while wearing full structural personal protective clothing.

Firefighters in another area called for assistance, and Lieutenant Rogers descended a set of stairs to join them. At the bottom of the stairs, Lieutenant Rogers told other firefighters that he did not feel well, sat down on a bench, and collapsed. Firefighters removed Lieutenant Rogers from the hazard area and handed him over to the care of EMS responders at the scene.

ALS-level care was provided on the scene and continued as Lieutenant Rogers was transported by ambulance to a local hospital. Emergency room staff at the hospital continued medical efforts, but Lieutenant Rogers was pronounced dead approximately an hour after becoming ill.

April 23, 2004 – Time Unknown
Alan David Toepke, Firefighter
Age 30, Wildland (Part-Time)
Midewin Agency Hotshot Crew, Illinois

Firefighter Toepke and the members of his hotshot crew were returning to their home base after assisting with firefighting efforts in the Apalachicola National Forest in Florida. The crew stopped in Arkansas to rest for the evening.

Witnesses saw Firefighter Toepke and two other firefighters running to cross Interstate 40. As the two leading firefighters entered the roadway, they were struck by a passing tractor-trailer truck. Firefighter Toepke was killed, the second firefighter was seriously injured, and the third firefighter was uninjured.

A law enforcement report on the incident indicated that the truck driver did not have time to react or avoid hitting the firefighters.

April 27, 2004 – 1741hrs
Jeffrey Charles Bergstrom, Firefighter/Paramedic
Age 34, Part-Time (Paid)
Stone Park Fire Department, Illinois

Firefighter/Paramedic Bergstrom was riding in the right front seat of an engine company responding to a structure fire.

An engine company from Northlake Fire Protection District was responding to the same structure fire. The Stone Park engine was northbound approaching an intersection, and the Northlake engine was eastbound approaching the same intersection.

The Northlake engine entered the intersection first, and the Stone Park engine struck the Northlake engine near the right rear tire. The force of the collision threw Firefighter/Paramedic Bergstrom forward, causing a fatal injury to his head. Firefighter/Paramedic Bergstrom was not wearing a seatbelt.

After the collision, the Stone Park engine entered into a clockwise rotation and the engine came to rest on its right side. Firefighter/Paramedic Bergstrom was ejected from the vehicle during the rotation. Responding firefighters and EMS personnel provided treatment on the scene, and Firefighter/Paramedic Bergstrom was transported to the hospital.

The driver of the Stone Park engine received severe injuries, the rear passenger received minor injuries, and the three occupants of the Northlake engine received minor injuries.

The intersection was controlled by an emergency vehicle preemption system that gives the green light to properly equipped response vehicles. A computer log of the operation of the intersection indicated that the Northlake engine had the green light at the time of the crash.

Firefighter/Paramedic Bergstrom was also a paramedic for the Chicago Fire Department.

∽∾∽

April 30, 2004 – 0905hrs
Irwin E. "Buzz" Gross, Firefighter
Age 58, Career
Brookline Fire Department, Massachusetts

Firefighter Gross and the members of his engine company were dispatched to a report of a gas odor in a structure. Firefighter Gross was a rear-seat passenger on the left side of the engine. Firefighter Gross was seen by other firefighters in a seated position and was engaged in preparing a gas meter for use on the incident. A witness to the incident told police that Firefighter Gross was standing inside the vehicle.

The engine made a right-hand turn from the fire station. As the apparatus entered the street, it flexed and the left-hand rear door opened. Firefighter Gross was ejected from the vehicle. He sustained a fatal head injury as a result of the fall. Firefighter Gross was transported to the hospital. He died on May 3, 2004, as a result of his injuries.

The apparatus was a 1976 Pircsh that served as a spare apparatus. The rear of the cab was equipped with doors. Police investigators found a seatbelt tucked under the seat used by Firefighter Gross; the seatbelt did not appear as if it had been used for quite some time. None of the four crew members on board the engine at the time of the incident was using a seatbelt.

In October 2004, a lawsuit was filed on behalf of the family of Firefighter Gross seeking damages from the Town of Brookline. There were also news reports of problems with the door latch on the rear door closest to Firefighter Gross.

~∞∞∞~

May 3, 2004 – 1908hrs
Grady Roy Austin, Captain
Age 74, Volunteer
Henderson County Fire Department, Tennessee

Captain Austin and the members of his fire department were conducting regular monthly training. The subject of the night's training was going to be pumping and drafting. There was not enough room in the fire station parking lot for the exercise, so the training was relocated to a school parking lot a few hundred feet from the fire station.

Some firefighters rode in apparatus from the station to the school parking lot, and others rode in personal vehicles. Captain Austin was a passenger on the tailgate of a pickup truck.

As the truck prepared for a turn, Captain Austin slipped from the tailgate and struck his head on the pavement. He received a severe head injury. Firefighters who witnessed the fall called for medical assistance and began treatment. CPR was initiated by firefighters and continued through the arrival of paramedics and transportation to the hospital.

Upon his arrival at the hospital, Captain Austin was pronounced dead as a result of his head injury.

~∞∞∞~

May 11, 2004 – 1035hrs
Joseph Edward "Pappy" Boles, Firefighter
Age 57, Part-Time (Paid)
West Area Volunteer Fire Department, Inc., North Carolina

Firefighter Boles had reported for the start of his shift at 0900hrs. He initially complained of a sharp pain in his left side rib area but said that it had gone away. Firefighter Boles assisted other firefighters with station and apparatus maintenance duties.

When firefighters took a break from their work, Firefighter Boles sat on a couch and took a drink of water. A few moments later, firefighters found Firefighter Boles on the couch slumped over and unresponsive.

continued on next page

65

Firefighter Boles was moved to the floor for treatment. CPR was initiated, and an AED was applied. Paramedics arrived and assisted with treatment. Firefighter Boles was transported by ambulance to the hospital. He was pronounced dead at the hospital after additional measures failed to revive him. The death was caused by a heart attack.

May 13, 2004 – 1700hrs
Randy Rayford Henderson, Forestry Technician
Age 42, Wildland (Full-Time)
United States Department of Agriculture Forest Service, Bienville Ranger District, Mississippi

Forestry Technician Henderson was detailed to the Osceola National Forest in Florida. He was assigned as the Forest's Fire Management Officer.

Forestry Technician Henderson and other firefighters were fighting the Mailbox Fire near Lake City, Florida. Forestry Technician Henderson was assigned as a Safety Officer on the incident.

The duties of the Safety Officer included monitoring firefighters, firefighting activities, and the weather. Forestry Technician Henderson was in radio contact with the IC.

After Forestry Technician Henderson did not respond to radio calls, a search was initiated. Forestry Technician Henderson was found lying on the ground facing the fire. He had sustained burns. His collapse, however, was caused by a fatal heart attack.

May 13, 2004 – 1752hrs
Jeffrey Warnick Howell, Firefighter
Age 42, Volunteer
Sharon Springs Fire Department, New York

Firefighter Howell and members of his fire department were fighting a lightning-caused fire in a residence. Firefighter Howell arrived ahead of responding fire apparatus and helped the occupants of the house remove personal belongings.

Upon the arrival of fire apparatus, Firefighter Howell donned his personal protective equipment (PPE) and assisted with the deployment of hoselines from an engine.

Firefighter Howell was not feeling well and sat down. The on-scene safety officer asked EMS responders to assess Firefighter Howell's condition. Firefighter Howell walked to the rear of a rescue truck, sat down, and collapsed.

CPR was initiated immediately, and Firefighter Howell was transported by ambulance to the hospital. Firefighter Howell was pronounced dead shortly after his arrival at the hospital. The cause of death was listed as a heart attack as a consequence of physical exertion in a hot, humid environment.

~~~

May 13, 2004 – 0730hrs
Harry Edward Suggs II, Fire Chief
Age 28, Volunteer
Green Pond Volunteer Fire and Rescue Service, Alabama

Chief Suggs was attending an EMS conference in Florida. He became ill and died at the conference. The death was as a consequence of taking prescribed medications for a recent surgery.

~~~

May 14, 2004 – 0608hrs
Michael Stokes Martin, Firefighter
Age 18, Volunteer
Ebenezer Volunteer Fire Department, South Carolina

Firefighter Martin was responding to an EMS incident in his personal vehicle, a 2004 Ford pickup truck.

As Firefighter Martin responded to the incident, the right wheels of the vehicle left the roadway. The vehicle came back onto the roadway and then left the roadway again on the left side, struck a culvert, and came to rest against a tree. Firefighter Martin was pronounced dead at the scene.

Firefighter Martin was not wearing a seatbelt, although an air bag was deployed as a result of the crash. The law enforcement crash report stated that the speed of the vehicle prior to the crash was 65 miles per hour in a 45 miles-per-hour zone.

~~~

May 17, 2004 – 1500hrs
Connie C. Bornman, Firefighter/EMT
Age 57, Volunteer
Middle River Volunteer Ambulance and Rescue Company, Maryland

Firefighter/EMT Bornman responded to an EMS incident. She assisted with patient treatment then went outside to get some fresh air. She sat in the front of the EMS vehicle with the air- conditioning running.

A family member of the original patient noticed that Firefighter/EMT Bornman was sick. Another firefighter went to check on her and found her unresponsive. She was treated at the scene and transported to a hospital in Baltimore.

continued on next page

Despite efforts at the scene, in the ambulance, and at the hospital, Firefighter/EMT Bornman died of a heart attack.

Firefighter/EMT Bornman was the first Baltimore County female firefighter to die in the line of duty.

June 5, 2004 – 1030hrs
Lawrence Joseph "Larry" Hoffman, Protection Unit Supervisor
Age 51, Wildland (Full-Time)
Oregon Department of Forestry

Protection Unit Supervisor Hoffman was completing the most arduous version of the annual recertification work capacity "pack" test. The test requires a hike of 3 miles with a 45-pound pack in less than 45 minutes.

About two-thirds of the way through the test, Protection Unit Supervisor Hoffman collapsed. Other firefighters taking the test and on standby for the test provided immediate medical assistance. Protection Unit Supervisor Hoffman was transported to the hospital but still succumbed to the cardiac-related illness.

June 8, 2004 – 1930hrs
William Richard Grudzinski, Assistant Chief
Age 46, Volunteer
Bridger Volunteer Fire Department, Montana

Assistant Chief Grudzinski and other members of his fire department were completing the low-intensity version of the annual work capacity "pack" test. The test requires a hike of 1 mile without a pack in less than 16 minutes.

After the completion of the test in 14 minutes and 1 second, Assistant Chief Grudzinski commented that the test had "kind of kicked his butt." Firefighters went back to the firehouse. While at the fire station, Assistant Chief Grudzinski was seen checking his pulse frequently. At the time, the frequent checks did not cause any concern.

Approximately 1 hour and 45 minutes after completing the test, Assistant Chief Grudzinski collapsed due to a heart attack. Firefighters who heard the dispatch rushed to Assistant Chief Grudzinski's home to assist EMS responders. Despite all efforts to revive him, Assistant Chief Grudzinski died.

June 17, 2004 – 1746hrs
Wayne Carl Turner, Pilot
Age 58, Wildland Contract
New Frontier Aviation under contract to the Bureau of Land
Management, Utah

Pilot Turner was operating a SEAT on a complex of wildland fires near
Brookside, Utah. Pilot Turner had just completed a fire retardant drop on
his third pass over the target area.

Immediately after the drop, the airplane's nose pitched down about
45 degrees. The aircraft maintained this attitude until it crashed. Pilot
Turner was killed instantly.

For additional information about this crash, consult the National
Transportation Safety Board Web site at http://www.ntsb.gov/ntsb/
query.asp - NTSB identification LAX04GA243.

June 17, 2004 – 1830hrs
Willie J. Lacy, Firefighter
Age 47, Career
Augusta Fire Department, Georgia

Firefighter Lacy had participated in training for most of the day and had
responded with the members of his engine company to several EMS calls.

After returning to the fire station, other firefighters discovered Firefighter
Lacy unconscious in the fire station dorm. Emergency medical
treatment was provided, and Firefighter Lacy was transported to the
hospital. Firefighter Lacy died of a cardiac-related illness.

June 19, 2004 – 1700hrs
Joshua St. Jermaine Martin, Junior Firefighter
Age 15, Volunteer
Duson Volunteer Fire Department, Louisiana

Junior Firefighter Martin was a front-seat passenger in a personally-
owned 2000 Ford Explorer being driven by another firefighter. The
firefighters were responding to an apartment fire and were operating
grille-mounted red lights.

The driver lost control of the vehicle, it rotated counter-clockwise,
crossed the center line of the roadway, and was struck on the right side
by a vehicle traveling in the opposite direction. The crash occurred in
daylight while light rain was falling. Water accumulation on the roadway
may have played a role in the loss of control.

Junior Firefighter Martin was killed in the crash. Neither the driver nor
Junior Firefighter Martin was wearing his seatbelt at the time of the crash.

June 19, 2004 – 1400hrs
Gary Dean Archibeque, Firefighter
Age 39, Paid-on-Call
Show Low Fire District, Arizona

Firefighter Archibeque and other firefighters were participating in a "Fuels Reduction" program sponsored by the Show Low Fire District. Residents in the wildland interface clear brush from their land, and the fire district sends a crew to chip and haul away the trees and brush that have been cleared.

Firefighter Archibeque was assigned as a crew leader for chipper operations. The crew began operations at 0700hrs. Firefighter Archibeque began to complain of indigestion later in the morning, and some homeowners gave him antacids. At approximately 1400hrs, as he was feeding tree limbs into the chipper, Firefighter Archibeque went to his knees. He signaled the crew to kill the chipper, members of his crew and bystanders helped him to the ground.

Firefighter Archibeque was checked and found to be pulseless. CPR was initiated and an ambulance was called. Firefighter Archibeque was defibrillated at least three times on scene and transported to the hospital. He was pronounced dead at the hospital. The cause of death was a heart attack.

June 22, 2004 – 2300hrs
Kenneth W. Lipyance, Lieutenant
Age 46, Volunteer
Churchill Volunteer Fire Company, Pennsylvania

Lieutenant Lipyance responded with members of his fire department to a motor vehicle crash that involved a piece of fire apparatus. Extrication was required at the scene, and Lieutenant Lipyance told other firefighters that he was not feeling well.

As the responders arrived back at their fire station, they came upon another motor vehicle crash a short distance from the station. Firefighters, including Lieutenant Lipyance, provided assistance on the scene of the second crash. Lieutenant Lipyance again told other firefighters that he was not feeling well, and then walked to the fire station.

Firefighters witnessed Lieutenant Lipyance drive slowly out of the fire station parking lot and then continue across four lanes of traffic until the vehicle hit the curb and came to rest. Lieutenant Lipyance was treated at the scene and then flown to a hospital by medical helicopter.

Lieutenant Lipyance died on June 30, 2004, as a result of a CVA and heart disease.

June 23, 2004 – 0927hrs
Thomas DeAngelis, Captain
Age 40, Volunteer
Independent Hose Company #2, Stowe Township, Pennsylvania

Captain DeAngelis responded with members of his fire department to an odor of gas report at 2120hrs on June 22, 2004. After sampling the air in the area, the fire chief discovered that a utility company crew was working on a gas line repair on a nearby street. After consulting with the repair crew, fire companies returned to the station and went into service at 2146hrs.

At 0218hrs on June 23, 2004, Captain DeAngelis' fire department was dispatched to a mutual-aid fire incident. Captain DeAngelis did not respond to the incident. Captain DeAngelis' lack of response was unusual. His wife later reported that he did not feel well enough at the time of dispatch to respond to the incident.

At 0830hrs on June 23, 2004, it was reported that Captain DeAngelis had suffered a serious heart attack and was at a local hospital. Prior to the arrival of firefighters at the hospital, Captain DeAngelis died.

July 9, 2004 – 1528hrs
Gary Don Fox, Fire Chief
Age 60, Volunteer
Bluegrove Volunteer Fire Department, Texas

The Bluegrove Volunteer Fire Department (VFD) received a report from a passerby of a burning vehicle in a hay field. Chief Fox responded alone to the call in a 1-ton brush truck carrying 300 gallons of water. Upon arrival at the fire at 1539hrs, the burning vehicle, a pickup truck with a round hay bale-carrying attachment, was well-involved. Chief Fox, assisted by the property owner, turned his attention to extinguishing the fire in the field. Fox was not wearing any firefighter protective equipment and was dressed in jeans and a long-sleeve shirt.

Mutual-aid assistance from the Henrietta Fire Department was requested at 1558hrs, the approximate time that the Bluegrove Volunteer Fire Department brush truck ran out of water. Upon arrival of the Henrietta units, Chief Fox left the scene, complaining that he had become overheated, and returned to the fire station with the Bluegrove VFD brush truck.

A Clay County deputy sheriff had spoken with Fox at the fire and described Fox as appearing pale and sweating profusely. Henrietta Fire Department units remained on the scene and extinguished the field and vehicle fire.

Bluegrove firefighters visited with Chief Fox at his home approximately 30 minutes after he returned the brush truck to the fire station. They described him as still appearing sweaty and hot. Fox declined any assistance and said he would wait for his wife's arrival. The firefighters

continued on next page

spoke with Fox's wife outside the home when she arrived 10 minutes later. She told them that Chief Fox had called her to come home because he had gotten overheated.

After Fox's wife entered the home, she observed Fox slumped on the couch, unresponsive, not breathing, and without a pulse. She called for help from the firefighters outside and called 9-1-1. Fox's wife (an EMT) and the firefighters initiated and continued CPR until the arrival of the ALS ambulance 15 minutes later. The ambulance took Fox to Clay County Medical Center where he was pronounced dead. The attending emergency room physician stated that Fox had a heart attack that could have been brought on from heat-related illness. No autopsy was ordered.

The summary above comes from a thorough report on this incident that was prepared by the Texas State Fire Marshal. The report is available at http://www.tdi.state.tx.us/fire/fmloddinvesti.html

July 11, 2004 – 2007hrs
Harold Dean Chappell, Firefighter
Age 53, Volunteer
Arlington Fire and Rescue, Inc., North Carolina

Firefighter Chappell responded with members of his fire department to a motor vehicle crash involving an overturned vehicle. While on scene, Firefighter Chappell assisted with traffic control and scene safety.

After the conclusion of the incident, Firefighter Chappell complained of not feeling well. He told another firefighter that he felt weak and that he was experiencing indigestion. Firefighter Chappell went home and took some antacids.

Firefighter Chappell rose the next day and was doing work at his residence when he collapsed. He was transported by ambulance to the hospital where he was pronounced dead.

The cause of death was listed as atherosclerotic cardiovascular disease.

July 13, 2004 – 0800hrs
Daniel Earl Elkins, Captain
Age 47, Career
Los Angeles County Fire Department, California

On July 12, 2004, Captain Elkins worked overtime at the Pine Fire Command Post (CP). When he was called in to work the fire, Captain Elkins drove to the CP in his personal vehicle. At 0715hrs on July 13, 2004, Captain Elkins was released from duty at Pine Fire and headed to his normal work assignment at Los Angeles County Fire Station 117.

As he drove to the fire station, Captain Elkins' vehicle drifted off of the road, struck a berm/culvert, went airborne for approximately 94 feet, and overturned several times before coming to rest. Although he was

72

wearing a seatbelt, and airbags did deploy, Captain Elkins was killed in the crash.

~~∿∿∿~~

July 21, 2004 – 0030hrs
Lester Phillips, Fire Chief
Age 72, Volunteer
Sunshine Volunteer Fire Department, Kentucky

Chief Phillips and the members of his department responded to a working fire in a manufactured home. The cause of the fire was arson. Chief Phillips complained of not feeling well at the scene. Firefighters helped Chief Phillips to his personal vehicle to get him away from the smoke. Chief Phillips drove back to the fire station and then to his home.

After consulting with family members, Chief Phillips consented to go to the hospital. Chief Phillips did not recover from his heart attack and died on August 4, 2004.

~~∿∿∿~~

July 29, 2004 – 1900
Thomas Conway, Fire Police Captain
Age 78, Volunteer
Haddon Heights Fire Department, New Jersey

Fire Police Captain Conway and members of his fire department responded to a carbon monoxide alarm activation. Fire Police Captain Conway directed traffic around responding fire apparatus. While on scene, he complained to other firefighters that he was not feeling well.

Fire Police Captain Conway declined an offer from other firefighters for an evaluation by EMS personnel; he said that he did not want to worry his wife. At the conclusion of the incident, Fire Police Captain Conway drove home.

Within minutes of arriving home, Fire Police Captain Conway called 911 and requested medical help. He was transported to the hospital by ambulance. After initial treatment at a local hospital, Fire Police Captain Conway was transferred to a regional hospital. His condition deteriorated and he died on August 2, 2004, of cardiac-related problems.

~~∿∿∿~~

July 31, 2004 – 1513hrs
George Henry Raber, Crew Chief
Age 69, Volunteer
Hebron Fire Protection District, North Dakota

Crew Chief Raber was operating a 2,500-gallon water tanker at the scene of a wildland fire. A hay bailer had started the fire. This was the second time that firefighters had responded to the scene that day.

continued on next page

When the engine running the pump on the apparatus stopped operating, Crew Chief Raber stopped the truck and set the parking brake. As he exited the truck, Crew Chief Raber fell to the ground and did not get up.

A farmer on the scene witnessed the fall, and firefighters immediately came to Crew Chief Raber's assistance. Crew Chief Raber was transported by ambulance to the hospital. He was pronounced dead at the hospital. The death was due to a heart attack.

August 4, 2004 – 2100hrs
Michael Benton McAdams, Firefighter
Age 69, Volunteer
Sapello-Rociada Volunteer Fire Company, New Mexico

On August 3, 2004, Firefighter McAdams and the members of his fire department had completed a regularly scheduled training session. The department was dispatched to a rollover vehicle crash about 15 miles from the fire station.

Firefighters assessed the injuries received by the driver of the vehicle, and found an intoxicated individual who was not in need of medical treatment. Law enforcement personnel transported the driver from the scene, and firefighters returned to service.

Firefighter McAdams went home and went to bed. He rose before 0800hrs the next morning and began to prepare breakfast. He complained to his wife of a sharp pain in his back and fell over. Family members, who were members of the fire department, started CPR, and they called an ambulance.

The ambulance broke down en route to the McAdams residence, so law enforcement personnel brought EMS personnel to the home. Firefighter McAdams was pronounced dead shortly after their arrival at the home. The cause of death was arteriosclerotic cardiovascular disease.

August 10, 2004 – 1730hrs
Barbara Louise Bordenkircher, Firefighter
Age 52, Volunteer
Wickliffe Rural Fire Department, Kentucky

Firefighter Bordenkircher was the driver of an engine apparatus responding to a wildland fire. The fire was caused when a tire came off of a boat trailer as it was being towed; the resulting sparks ignited nearby brush.

74

After the right wheels of the apparatus left the roadway, Firefighter Bordenkircher steered left to bring the truck back on the road and the then lost control of the vehicle. The engine crashed into a tree, and Firefighter Bordenkircher was ejected.

Another firefighter who was a passenger in the engine was severely injured. The law enforcement report on the crash cited a shift in the water load as the cause of the incident.

~~~

August 11, 2004 – 1025hrs
Mike Ward, Pilot
Age 55, Wildland Contract
Shasta Aviation under contract to the United States Forest Service, Washington

Pilot Ward was ferrying supplies to firefighters working in the Alpine Lakes Wilderness. The supplies were carried below the helicopter in a sling. There were loads for two drop sites in the sling.

The helicopter arrived at the first site and lowered the sling to the ground. Firefighters released the load intended for the first site and reattached the second load meant to be dropped at the other site.

The helicopter rotated as it began to leave the first drop site, and the aircraft's tail rotor contacted a dead tree. The helicopter began to spin and fell to the ground. As the helicopter fell, the aircraft began to disintegrate, and a fire erupted. Pilot Ward was killed in the crash.

For additional information about this crash, consult the National Transportation Safety Board Web site at http://www.ntsb.gov/ntsb/query.asp - NTSB identification SEA04TA158.

~~~

August 14, 2004 – 1342hrs
Jaime Leah Foster, Firefighter
Age 25, Career
Los Angeles City Fire Department, California

Firefighter Foster and the members of her company responded to a working fire in a residence. Firefighter Foster was working overtime at her regularly assigned fire station. Once firefighting operations were completed, Firefighter Foster's company was released from the scene. In order to leave the scene, Firefighter Foster's apparatus had to back out of a side street onto another street.

Firefighter Foster took a position on the tailboard, or back step, of the engine near the buzzer used to signal the apparatus driver. The Company Officer (CO) was behind the apparatus in view of the driver's rearview mirror. Upon receiving the standard signal from Firefighter Foster, the apparatus driver began to back up at a speed estimated at 2-1/2 miles per hour.

continued on next page

The CO turned away to control traffic as the apparatus neared the intersection. When the officer turned back to look at the engine, Firefighter Foster was no longer seen on the back of the apparatus. The officer ran toward the apparatus and, upon seeing Firefighter Foster on the ground, yelled for the apparatus driver to stop.

Firefighters on the scene quickly ran to aid Firefighter Foster. An ambulance that had responded to the original incident returned to the scene and transported Firefighter Foster to the hospital. A total of 12 minutes passed between Firefighter Foster's injury and her arrival at the hospital.

Firefighter Foster was pronounced dead at the hospital. The cause of death was listed as multiple blunt trauma injuries.

⁓⁓⁓

August 16, 2004 – Time Unknown
Robert M. Weber, Jr., Corporal
Age 23, Career
Marine Corps Air Station Beaufort, South Carolina

Corporal Weber was on duty and riding in an Aircraft Rescue Firefighting (ARFF) vehicle. Corporal Weber and the other members of his crew were standing by near a runway as aircraft operations occurred.

Corporal Weber collapsed suddenly. Firefighters provided medical treatment, including the use of an AED. Corporal Weber was transported to the hospital where he was pronounced dead. The death was due to a heart attack.

⁓⁓⁓

August 20, 2004 – 2020hrs
John Daniel Taylor, Jr., Captain
Age 53, Career
Philadelphia Fire Department, Pennsylvania

Rey Rubio, Firefighter
Age 42, Career
Philadelphia Fire Department, Pennsylvania

Captain Taylor and Firefighter Rubio were members of an engine company crew that was dispatched to a fire in a residential structure. Upon arrival on the scene, a working fire was found.

Captain Taylor, Firefighter Rubio, and another firefighter deployed an attack line into the basement of the structure. While in the basement, conditions deteriorated rapidly, and Captain Taylor ordered his crew to evacuate the basement.

As the crew began its withdrawal, Firefighter Rubio's SCBA became entangled. He was unable to leave. Captain Taylor ordered the third firefighter, a rookie, to leave as he attempted to free Firefighter Rubio.

76

Rescue efforts to assist the trapped firefighters were delayed by the volume of fire. Once the fire was knocked down, firefighters entered the basement and removed Captain Taylor and Firefighter Rubio. Despite their best efforts, Captain Taylor and Firefighter Rubio died of smoke inhalation.

Electrical wiring being used to provide lighting for an illegal marijuana-growing operation caused the fire. A man was charged with murder as a result of the firefighter's deaths.

~~~

August 22, 2004 – 1427hrs
Robert E. Woolf, Firefighter
Age 63, Volunteer
Phillipsburg Fire Department, Ohio

Firefighter Woolf and members of his fire department were in the process of cleaning up after a fundraising event. The fire department had borrowed tables and chairs from a local community center, and firefighters had returned a number of them in one firefighter's personal pickup truck.

Firefighter Woolf and another firefighter were riding back to the fire station on the lowered tailgate of the pickup truck. The straps that hold the tailgate in position failed and both firefighters fell to the ground while the vehicle was in motion.

Another firefighter driving a vehicle following the pickup witnessed the fall and stopped immediately. Other firefighters began medical treatment immediately; Firefighter Woolf was motionless and unresponsive. He was transported quickly to the hospital.

Hospital staff performed surgery on Firefighter Woolf the same day as the fall. He did not recover and died of his head injury on August 25, 2004. The injuries received by the other firefighter were non-life-threatening.

~~~

August 23, 2004 – 1725hrs
Benjamin Matthew Lang, Firefighter/EMT
Age 22, Career
Polk County Fire Department, Florida

Firefighter Lang was working an overtime shift for the fire department and was a passenger in the rear of an ambulance transporting a patient to the hospital. The ambulance was traveling in an emergency mode, using emergency lights and the siren.

The road was wet from recent rains, and there still was some shower activity. The driver of the ambulance lost control; the ambulance left the roadway, slid into a grass area, overturned, and struck a large tree. The impact with the tree inflicted major damage on the patient compartment of the ambulance.

continued on next page

Firefighter/EMT Lang was severely injured and trapped in the wreckage of the ambulance. His death was caused by blunt force trauma to the head.

Firefighter Lang was sitting in the rear-facing seat at the patient's head at the time of the crash. It is unknown if Firefighter Lang was wearing a seatbelt.

~~~

August 25, 2004 – 1815hrs
David Edward Vinisky, Firefighter
Age 49, Volunteer
Raccoon Township Independent Volunteer Fire Department #1, Pennsylvania

Firefighters had completed an evening of training on a new pumper that had been delivered recently. Firefighter Vinisky and another firefighter were photographing the apparatus.

Firefighter Vinisky and the other firefighter walked to the rear of the apparatus. At the same time, another firefighter entered the cab of the apparatus and began to back the truck into the fire station. Both firefighters at the rear of the truck were knocked off of their feet. The second firefighter managed to grab the rear of the apparatus and was dragged along under the vehicle. This firefighter yelled for the driver to stop the vehicle.

Hearing the yelling from the firefighter under the rear of the apparatus, the driver stopped, placed the vehicle in forward gear, and pulled forward a few feet. Firefighter Vinisky was crushed under the wheels of the apparatus as it backed up and then crushed again as the apparatus pulled forward.

Firefighters and EMS personnel responded to the scene. Firefighter Vinisky was pronounced dead at the scene due to massive trauma.

The driver of the pumper did not have a valid operator's license. He was charged by law enforcement officials with causing the accident; however the charges were dropped at the request of the fire department and the Vinisky family.

~~~

August 28, 2004 – 1600hrs
Cordell W. "Cory" French, Firefighter/EMS Director
Age 44, Volunteer
Towanda Fire/Rescue, Kansas

Firefighter French and the members of his fire department participated in a drill involving hoseline usage, map training, and vehicle maintenance. The drill was completed at approximately 1600hrs.

Firefighter French returned home and suffered a heart attack at approximately 2000hrs. Fellow firefighters responded and provided treatment for Firefighter French. Despite their efforts, Firefighter French was pronounced dead at 2130hrs as the result of a heart attack.

September 3, 2004 – 0304hrs
James J. D'heron, Deputy Chief
Age 51, Career
New Brunswick Fire Department, New Jersey

Deputy Chief D'heron and members of his fire department were dispatched to a fire in a multiple-family residence. Deputy Chief D'heron arrived first, reported a working fire, and took command. Not seeing any occupants outside or leaving the structure, he entered the structure to alert residents to the fire. Deputy Chief D'heron was not wearing any personal protective clothing or equipment.

Deputy Chief D'heron was banging on doors to alert residents of the fire when an explosion occurred. Deputy Chief D'heron was mortally burned in the explosion and ensuing fire. Arriving firefighters found him on the second-floor landing and removed him from the structure. He was pronounced dead at the scene.

A homeless man who slept in the hallway of the building caused the fire. The man discarded a cigarette near some plastic shower curtains and a plastic container of gasoline. The curtains and the gasoline ignited, producing sufficient heat to cause three nearby propane cylinders to vent their contents. The accumulated propane gas is thought to be the source of the explosion that killed Deputy Chief D'heron.

September 5, 2004 – 1715hrs
Gerald Kerr "Mac" McGowan, Acting Captain
Age 57, Career
Kansas City Fire Department, Missouri

Acting Captain McGowan and the members of his engine company were responding to an apartment fire in their first-due area. The company was responding with emergency lights and siren in operation on a four-lane highway.

Traffic in the right-hand lane stopped to yield for the responding engine, and a car stopped in the left-hand lane. As the engine crossed the center line to go around the stopped traffic, the car in the left-hand lane attempted to execute a left-hand turn. The engine veered to the left in an attempt to avoid a collision, struck the turning vehicle, struck a vehicle parked in the opposing direction of traffic, struck a utility pole, and then struck a large tree. The area where Acting Captain McGowan was seated bore the brunt of the collision.

continued on next page

Acting Captain McGowan was extricated and transported to a hospital. He was unresponsive when first assessed and never regained heart functions. He was pronounced dead at the hospital as a result of his injuries.

The driver of the turning vehicle was operating without a valid driver's license. Acting Captain McGowan was not wearing a seatbelt, the apparatus driver was wearing a seatbelt, and the seatbelt usage of the two passenger firefighters was unknown.

～～～

September 9, 2004 – 0130hrs
Steven M. Rosenfeld, EMS Captain
Age 52, Volunteer
Salem Volunteer Fire Department, Virginia

Captain Rosenfeld and the members of his fire department responded to a motor vehicle crash involving injuries. Captain Rosenfeld had just finished loading two patients into ambulances when he collapsed.

Emergency responders on the scene provided immediate ALS-level care and Captain Rosenfeld was transported to the hospital. He was pronounced dead as the result of a heart attack approximately 90 minutes after becoming ill.

～～～

September 10, 2004 – 2145hrs
Richard O'Brien, Firefighter
Age 63, Volunteer
Warren Fire Department, Rhode Island

Firefighter O'Brien and the members of his fire department responded to a fire in an apartment building. Firefighters found a fire in the kitchen of a third-floor apartment. Firefighter O'Brien provided ventilation in the fire occupancy by opening windows.

Firefighter O'Brien collapsed and was transported by fire department rescue unit to the hospital. He was pronounced dead later that night as the result of a heart attack.

The fire occurred in a building owned, but not occupied, by Firefighter O'Brien.

～～～

September 12, 2004 – 1500hrs
Eva Marie Schickee, Firefighter
Age 23, Wildland (Full-Time)
California Department of Forestry and Fire Protection

Firefighter Schickee was a member of an elite nine-person helitack team. The team was called to fight the beginning stages of the Tuolumne Fire in the Stanislaus National Forest. When the team arrived

80

aboard their helicopter, seven members of the team were dropped off, and two members remained with the helicopter to begin water drops.

Firefighters established safety zones and began their descent into a steep canyon. The fire was moving up the canyon away from them and had displayed mild behavior at this point in the incident. While the crew was working, a wind shift blew the fire toward the crew.

With only moments to react due to the speed of the fire, firefighters were forced to run to their safety zones. Firefighter Schickee and another firefighter attempted to run uphill to the roadside safety zone. The first firefighter made it to safety; the fire overran Firefighter Schickee before she could reach the safety of the road.

The cause of death for Firefighter Schickee was smoke inhalation. Firefighter Schickee was the first female CDF firefighter to be killed in the line of duty.

September 14, 2004 – Time Unknown
Kevin L. Slain, Captain
Age 47, Career
Dixon Rural Fire Protection District, Illinois

Captain Slain attended a regular meeting of the fire protection district's board of directors. After the meeting, firefighters were standing in the apparatus bay talking. Captain Slain told others that he was not feeling well.

Captain Slain walked to the ambulance located in the station and was treated and transported to the hospital by firefighters. Captain Slain was pronounced dead at 2116hrs as the result of a cardiac-related illness.

September 16, 2004 – 1934hrs
Clinton L. Romine, Firefighter
Age 25, Volunteer
Good Springs Volunteer Fire Department, Alabama

Firefighter Romine was working with the members of his fire department and others to remove fallen trees from roadways in the wake of Hurricane Ivan.

Firefighter Romine went to a merchant to purchase a replacement chain for his personal saw. As he drove back to the fire station, a portion of a storm-damaged tree fell onto his personal vehicle. The fallen tree was approximately 2-1/2 feet in diameter and fell directly on the cab of Firefighter Romine's truck. He was killed instantly. A passenger in the truck received non-life-threatening injuries.

September 17, 2004 – Time Unknown
John A. Brenckle, Fire Police Captain
Age 57, Volunteer
Berkeley Hills Fire Company, Pennsylvania

Fire Police Captain Brenckle was assigned to the Ross Township Fire Police during the severe flooding that resulted from Hurricane Ivan. From September 17 to 19, Fire Police Captain Brenckle spent a great deal of time in and around water as he performed his fire police duties. The water was contaminated with unknown types and quantities of materials, including chemicals and human waste due to overwhelmed sewer systems.

Fire Police Captain Brenckle became ill and was admitted to a local hospital. Two days after being admitted to the hospital, Fire Police Captain Brenckle died. The death was the result of a bloodborne infection and existing diabetes. Contributing factors cited in relation to the death were hypertensive cardiovascular disease and exposure to flood water.

September 18, 2004 – 0850hrs
William Jim Lightbody, Rescue Member/Firefighter
Age 46, Volunteer
Paramus Volunteer Rescue Squad, Inc., New Jersey

Rescue Member Lightbody and members of his rescue company had completed their response to an entrapment incident on the Garden State Parkway.

As the apparatus returned to the station, Rescue Member Lightbody suffered a heart attack. He was taken to the hospital by ambulance but did not survive.

September 26, 2004 – 1126hrs
William Niles "Bill" Weborg, Assistant Fire Chief
Age 42, Volunteer
Ephraim Fire Department, Wisconsin

Assistant Chief Weborg and the members of his fire department were dispatched to a report of a boat fire. Upon arrival at the fire station, Assistant Chief Weborg collapsed of an apparent heart attack.

Firefighters provided medical assistance, and paramedics treated and transported Assistant Chief Weborg to the hospital. He was pronounced dead at the hospital a short time later.

October 2, 2004 – 1246hrs
Daniel Paul Holmes, Forestry Technician - Hotshot
Age 26, Wildland (Part-Time)
National Park Service, Sequoia and Kings Canyon National Parks,
California

Forestry Technician Holmes was a member of the Arrowhead Interagency Hotshot Crew. He was working as a saw team member cutting dead trees in an unburned section of a prescribed burn in the Kings Canyon National Park.

A dead tree near the perimeter of the controlled burn area was burning near its top, and falling embers from the tree threatened to spread the fire past the perimeter. Forestry Technician Holmes and the members of his saw team were assessing how best to drop the tree and prevent the spread of the fire.

Firefighters watching the team and the tree noticed that a section of the tree was breaking free and starting to fall. These firefighters called to Forestry Technician Holmes and his team and told them of the danger. Forestry Technician Holmes reacted immediately but was only able to take a couple of steps before being struck by the falling tree.

Other firefighters provided medical assistance to Forestry Technician Holmes, and an ambulance was used to transport him to a helicopter landing zone for evacuation. Forestry Technician Holmes went into cardiac arrest during the ambulance ride to the landing zone. After consulting with medical control, he was pronounced dead at the helicopter landing zone. The cause of death was due to head injuries.

October 7, 2004 – 1230hrs
Michael J. Kilpatrick, Assistant Fire Chief
Age 58, Volunteer
North Lake Fire Department, Wisconsin

Assistant Chief Kilpatrick and other firefighters were standing by at a fire station in a neighboring community during a major fire incident. Assistant Chief Kilpatrick became ill and collapsed.

EMS personnel standing by in the same fire station transported Assistant Chief Kilpatrick to the hospital. Despite the efforts of firefighters, EMS personnel, and hospital personnel, Assistant Chief Kilpatrick died. The cause of death was a heart attack.

October 10, 2004 – 1545hrs
Frederick A. Smith II, Firefighter
Age 33, Volunteer
Salem Center Volunteer Fire Department, Indiana

Firefighter Smith was working on horseback as a first responder at a horse ride for charity event. Firefighter Smith was riding his horse back to the lodge when his foot and stirrup got caught up in another horse, and he fell from his horse.

Firefighter Smith received serious head injuries in the fall. He was transported by helicopter to the hospital where he died a short time after the fall.

October 11, 2004 – 1810hrs
Steven Charles Brack, Firefighter
Age 36, Volunteer
Allentown Volunteer Fire Department, Georgia

Firefighter Brack was responding from his home to a motor vehicle crash that required extrication. He was driving his personal vehicle.

Less than one-fourth mile from his home, Firefighter Brack lost control of his vehicle. The car left the roadway and struck a culvert. The car then became airborne, flipped several times, and came to rest. Firefighter Brack was killed.

The streets were wet, and it was raining at the time of the incident.

October 15, 2004 – 0050hrs
William E. Bierbower, Firefighter
Age 72, Volunteer
Fairmont-Hahntown Volunteer Fire Department, Pennsylvania

Firefighter Bierbower responded to his fire station for a mutual- aid response. The response was cancelled by the requesting department prior to the departure of any apparatus. Firefighter Bierbower was speaking with the person completing the paperwork associated with the incident when he suddenly collapsed.

Firefighters began CPR, and an ambulance was called. Firefighter Bierbower was transported by ambulance to the hospital but was not revived. The cause of death was listed as hypertension and arteriosclerotic cardiovascular disease.

October 16, 2004 – 2015hrs
Jordan Lee Nonnemaker, Firefighter
Age 18, Volunteer
Amity Fire Company, Pennsylvania

Firefighter Nonnemaker was in his fire station with other firefighters. He left the station with another firefighter to rent movies to watch at the fire station.

Firefighter Nonnemaker was the front-seat passenger in a car driven by another firefighter. The firefighter lost control of the vehicle. It crossed the center line of the road, and an oncoming vehicle struck the passenger side of the firefighter's vehicle. Firefighter Nonnemaker was killed instantly.

October 20, 2004 – 0043hrs
Robert D. Heighton, Firefighter/Paramedic
Age 45, Career
South Walton Fire District, Florida

Firefighter Heighton was a member of the three-person medical helicopter crew. The South Walton Fire District staffed a paramedic spot on the helicopter through a joint airmedical service agreement with the Sacred Heart Health System.

Firefighter Heighton was on duty when his helicopter was dispatched for an interfacility transfer of a patient. After a weather check by the pilot, the helicopter lifted off at 0041hrs. At 0043hrs, the pilot called his dispatcher and reported that they were returning to base due to weather. The dispatcher did not receive any further contact from the helicopter and, assuming the aircraft had arrived back safely, cleared the call from the dispatch computer.

At 0610hrs, the pilot for the oncoming crew noticed that the helicopter was not present. The pilot spoke to the dispatch center, and a search for the helicopter was initiated. The wreckage of the helicopter was located at about 0820hrs. The wreckage was in about 10 feet of water.

The body of Firefighter Heighton was the last to be found. It was located on October 21, 2004 at 1120hrs. The cause of death was listed as multiple trauma.

For additional information about this crash, consult the National Transportation Safety Board Web site at http://www.ntsb.gov/ntsb/query.asp - NTSB identification MIA05FA008.

October 20, 2004 – 0930hrs
Mark A. Parrish, Deputy Chief
Age 50, Volunteer
Normandy Fire Protection District, Missouri

Deputy Chief Parrish was attending a meeting in the department's fire station. During the meeting, he fell from a chair and suffered a seizure later attributed to a fatal heart attack. He was treated and transported to the hospital by firefighters but did not survive.

October 20, 2004 – 1806hrs
Gary Allen Tilton, Fire Chief
Age 58, Career
Katy Fire Department, Texas

The Katy Fire Department units were dispatched to a crash involving multiple vehicles and injuries. Chief Tilton arrived on the scene and assisted with traffic control and other on-scene operations. The injured were treated and transported, and firefighters stood by for the arrival of law enforcement and towing services.

The IC observed that Chief Tilton looked fatigued and asked him if he was feeling well. The chief responded that it seemed to be taking longer and longer to recover from the treatments that he was receiving for an illness. After a short break, the chief seemed to be feeling better. When the incident was terminated, fire companies went back into service at approximately 1923hrs and Chief Tilton departed the accident scene.

At 2234hrs, firefighters were dispatched to Chief Tilton's residence for a medical emergency. Chief Tilton had suffered a heart attack. Firefighters provided ALS-level EMS care and transported Chief Tilton to the hospital. Despite their efforts, Chief Tilton was pronounced dead shortly after arriving at the hospital.

November 1, 2004 – 0956hrs
Lewis Ray "Lewie" McNally, Assistant Fire Chief
Age 42, Volunteer
Newmanstown Volunteer Fire Company, Pennsylvania

Assistant Chief McNally and other members of his fire department were dispatched to a vehicle crash as mutual aid. The Newmanstown rescue truck was cancelled during its response, but Assistant Chief McNally reached the scene and assisted with patient treatment.

Assistant Chief McNally left the crash scene, completed some personal business, and went home. He collapsed and died of a heart attack approximately an hour after the initial response to the vehicle crash.

November 2, 2004 – 2005hrs
Donald Nathan Carlson, Firefighter
Age 60, Volunteer
Ute Fire Department, Iowa

Firefighter Carlson and two other firefighters were in a fire department equipment van responding to the scene of a rollover motor vehicle crash. As the vehicle responded, firefighters noticed that Firefighter Carlson was leaning against a door and then dropped into the stepwell of the door.

The vehicle was stopped, firefighters removed Firefighter Carlson from the truck, and an ambulance was summoned. CPR and ALS-level treatment were provided, and Firefighter Carlson was transported to the hospital.

Despite all efforts, Firefighter Carlson died of a heart attack.

November 6, 2004 – 1809hrs
Charles Colbert "Littleman" Webb, Lieutenant
Age 63, Volunteer
Mayking Volunteer Fire Department, Kentucky

Lieutenant Webb was a passenger in a pumper responding to the report of a forest fire. The driver pulled the apparatus off of the road when he was told that Lieutenant Webb was ill.

Firefighters treated Lieutenant Webb for an apparent heart attack, and he was transported to the hospital. ALS-level treatment was provided by paramedics. He was pronounced dead at the hospital after all treatment efforts failed to revive him.

November 13, 2004 – 1900hrs
Edward George Schnauss, Captain
Age 54, Volunteer
Morrison Volunteer Fire Department, Missouri

Captain Schnauss was performing routine maintenance on fire apparatus inside the fire station. When he failed to contact family members, his wife went to the fire station to look for him and found him deceased underneath a fire engine.

Captain Schnauss had suffered a fatal heart attack.

November 15, 2004 – 0900hrs
Donald H. Kersting, Battalion Chief
Age 47, Career
Wichita Fire Department, Kansas

Chief Kersting arrived at work in the fire department administrative offices and went through his regular morning routine. At approximately 0900hrs, firefighters walking by Chief Kersting's office found him

continued on next page

slumped over his desk and unresponsive. An ambulance was called, and CPR was initiated. An AED was used in an attempt to revive Chief Kersting.

Chief Kersting died as the result of his illness. He was also a member of his hometown volunteer fire department.

November 15, 2004 – 2120hrs
Patrick J. Cramer, Firefighter
Age 51, Career
Chicago Fire Department, Illinois

Firefighter Cramer worked an overtime shift from 1000hrs on November 14, 2004 through 0800hrs on November 15, 2004. During the shift, Firefighter Cramer responded to eight emergency incidents, including a working structural fire.

At approximately 2120hrs on November 15, 2004, Firefighter Cramer was playing basketball with friends at a local high school gym. He suddenly collapsed of an apparent heart attack.

Friends called 9-1-1, and fire department paramedics responded. Firefighter Cramer was transported to the hospital but did not recover.

November 30, 2004 – 1009hrs
Jackson Huber "Jack" Gerhart, Firefighter
Age 65, Volunteer
Chambersburg Fire Department, Pennsylvania

Firefighter Gerhart and the members of his fire department were dispatched to a fire in a residence. Heavy smoke was showing as Firefighter Gerhart arrived in his personal vehicle before responding fire apparatus.

Firefighter Gerhart, knowing the route that responding fire apparatus would take to the scene, stood at a fire hydrant awaiting the arrival of the first-due engine company. When the engine arrived, Firefighter Gerhart motioned to the driver and indicated that he would advance the supply line to the hydrant.

Firefighter Gerhart mounted the back step of the apparatus and grabbed a 5-inch supply line. He fell backwards, still holding the hose, and struck the ground. The driver of the engine, not seeing Firefighter Gerhart advancing the line to the hydrant, got out of the engine to investigate. He found Firefighter Gerhart on his back and unconscious.

Firefighters came to the aid of Firefighter Gerhart. He was able to open his eyes but did not speak. He was transported by ambulance to a local hospital and then transferred by helicopter to a regional hospital.

Firefighter Gerhart remained in a coma until his death on December 5, 2004. The cause of death was listed as blunt force trauma to the head as a result of the fall.

Firefighter Gerhart was a life member of the Junior Hose and Truck Company, a volunteer company that is a part of the Chambersburg Fire Department, and had a 32-year career with the District of Columbia Fire Department and the International Association of Firefighters.

December 13, 2004 – 1740hrs
James Larry Rogers, Fire Chief
Age 55, Career
Claxton Volunteer Fire Department, Georgia

Chief Rogers was driving a fire department rescue truck back to the station from the scene of a working fire in a commercial chicken house. This incident was the third response of the day for Chief Rogers and his department.

Another firefighter following the apparatus witnessed it leave the road to the left, travel through a ditch, and come to rest. While still on the scene of the crash, Chief Rogers collapsed of an apparent heart attack.

Treatment was provided on the scene, but Chief Rogers was pronounced dead at 1815hrs.

December 14, 2004 – 1300hrs
William Jess "Bill" Briggs, Captain
Age 52, Career
INEEL Fire Department, Idaho

Captain Briggs complained of not feeling well while on duty on December 13, 2004. He suffered a heart attack while performing physical fitness activities the next day and died.

December 14, 2004 – 0224hrs
Michael R. Dunlap, Firefighter
Age 47, Volunteer
Hydetown Volunteer Fire Department, Pennsylvania

Firefighter Dunlap responded to five emergency incidents during a 9-hour period during a heavy snow storm. The last of these incidents was a mutual-aid EMS response. During the work at this incident, Firefighter Dunlap told other firefighters that he felt tired but did not complain of any other medical conditions.

continued on next page

Later in the day, Firefighter Dunlap called a relative to say that he was having chest pains. He was found a short time later, having suffered a heart attack. He was transported to the hospital but was not revived.

～∂∂～

December 17, 2004 – 1700hrs
Herbert C. "Herbie" Caldwell III, Firefighter
Age 49, Volunteer
Newberry Township Fire Department, Pennsylvania

Firefighter Caldwell and a passenger were responding in Firefighter Caldwell's personal vehicle to a report of smoke in a residence. As they drove, Firefighter Caldwell began to cough up blood. He pulled the car to the side of the road, and the passenger summoned assistance using a fire department radio. Despite efforts at the scene, in transit, and at the hospital, Firefighter Caldwell died.

Firefighter Caldwell suffered from a thoracic aneurysm, the failure of a blood vessel.

～∂∂～

December 20, 2004 – 1519hrs
Nito Rene Guajardo, Firefighter
Age 24, Career
Baytown Fire & Rescue, Texas

Firefighter Guajardo and the members of his quint company responded to a fire in a residential structure. Upon arrival on the scene, they found a working fire.

Firefighters, including Firefighter Guajardo, advanced an attack line into the interior of the structure. When conditions deteriorated, firefighters withdrew from the structure and conducted an accountability check. Firefighter Guajardo was found to be missing.

Firefighters could not immediately reenter the structure due to fire conditions. Once conditions allowed, Firefighter Guajardo was discovered near the front door of the residence and removed to the exterior. Firefighter Guajardo had been out of contact for 19 minutes prior to being discovered. Treatment was initiated but was not successful in reviving Firefighter Guajardo. Asphyxiation and burns caused his death.

A report on this incident will be prepared by the Texas State Fire Marshal. The report will be available at http://www.tdi.state.tx.us/fire/fmloddinvesti.html

～∂∂～

90

December 20, 2004 – 1657hrs
Theodore Arthur "Ted" Myhre, Firefighter
Age 73, Volunteer
Bishop Hill Fire Department, Illinois

Firefighter Myhre was a passenger in a fire department vehicle as it responded to a motor vehicle crash. Upon arrival on the scene, law enforcement personnel told firefighters that their services were not needed and released them from the incident.

The pumper turned into a driveway in order to turn around and return to station. Firefighter Myhre got out of the apparatus to control traffic. Firefighter Myhre was struck by a passing pickup truck as he stood in the roadway. Firefighter Myhre was transported to the hospital but died of his injuries. He was pronounced dead at 1748hrs.

Firefighter Myhre was carrying a flashlight but was not wearing protective clothing or reflective materials. The driver of the vehicle that struck Firefighter Myhre said that he saw Myhre too late to avoid hitting him, despite braking and swerving to avoid contact. The driver was not charged.

December 23, 2004 – 1930hrs
Jason Todd Rowe, Fire Chief
Age 30, Volunteer
Elkhorn City Fire Department, Kentucky

Chief Rowe was conducting a training session on the use of airbags to lift a fire truck to allow the installation of snow chains. An airbag dislocated and caused the truck to roll forward. A wheel block securing the front tire was also dislocated. Chief Rowe was under the rear of the apparatus when these actions occurred.

Firefighters asked Chief Rowe if he was alright and did not receive a response. They found Chief Rowe unconscious and suffering from a severe head injury and blood loss.

Chief Rowe was pronounced dead at a local hospital approximately 20 minutes after the incident.

December 25, 2004 – 0003hrs
John M. "Jackie" Stoudt, Chief Engineer
Age 66, Volunteer
Diligence Fire Company #1, Summit Hill, Pennsylvania

Chief Engineer Stoudt and the members of his fire department responded to three emergency incidents on December 24, 2004. Just after midnight on Christmas morning, firefighters were dispatched to a structure fire.

91

Chief Engineer Stoudt responded to the fire station and was climbing aboard a ladder truck when he fell backwards and hit the floor. Firefighters came to his aid, and he was soon flown by helicopter to a regional hospital. Chief Engineer Stoudt was pronounced dead as the result of a heart attack approximately 2 hours after collapsing.

December 27, 2004 – 2255hrs
James Harrison Fugate, Jr., Firefighter
Age 20, Volunteer
Collinsville Fire Department, Oklahoma

Firefighter Fugate was responding in his personal vehicle to a fire alarm in a residence. Firefighter Fugate lost control of the vehicle as it crested a hill. The vehicle skidded 197 feet, went off of the left-hand side of the roadway, flipped twice, and came to rest 138 feet from the point of first contact with the ditch.

Firefighter Fugate was not wearing a seatbelt at the time of the crash, and he was ejected during the first rollover. He was transported to the hospital by helicopter but died on January 11, 2005 as a result of his injuries.

The law enforcement report on the crash cited unsafe speed as a factor in the incident. The report estimated the speed of the vehicle at the time of the crash at 80 miles per hour in a 50 mile-per-hour zone.

December 29, 2004 – 0003hrs
Jared Michael Moore, Firefighter
Age 19, Volunteer
Fairmont Township Fire Department, Kansas

Firefighter Moore was responding in his personal vehicle to the scene of a single-vehicle rollover motor vehicle crash. The emergency flashers on Firefighter Moore's vehicle were in operation. Firefighter Moore slowed to begin a left-hand turn onto the road where the incident was located and was struck from behind by a sheriff's cruiser responding to the same incident.

Firefighter Moore was not wearing his seatbelt and was ejected from his vehicle. Firefighter Moore was unconscious at the scene and was flown by medical helicopter to a regional hospital. He was pronounced dead after arriving at the hospital.

The deputy sheriff driving the cruiser that struck Firefighter Moore's vehicle was charged with vehicular homicide. A law enforcement traffic investigation placed the cruiser's estimated speed at between 84 and 87 miles per hour at the time of the crash. According to a news account, the deputy was confused as to the location of the original crash and was in the process of passing Firefighter Moore.

INCIDENTS PRIOR TO 2004

March 15, 2003 – 1457hrs
Lawrence J. Sweetnich, Firefighter
Age 51, Career
Garfield Heights Fire Department, Ohio

Firefighter Sweetnich was the officer in charge at his fire station. Training on the station's quint apparatus was conducted including setup, climbing the aerial ladder, and pump operations.

Approximately 30 minutes after training was completed, Firefighter Sweetnich experienced upper abdominal pain. The pain subsided quickly. Approximately 15 minutes later, Firefighter Sweetnich went out the rear of the fire station to meet his wife in the parking lot.

As Firefighter Sweetnich walked across the parking lot, he collapsed suddenly. His wife called for help, and firefighters and EMS workers responded from inside the fire station. Firefighter Sweetnich was treated on the scene, during the trip to the hospital, and upon his arrival at the emergency room. He was later pronounced dead as a result of a heart attack.

July 4, 2003 – 2140hrs
Raymond L. Peterman, Fire Captain
Age 62, Career
Los Angeles City Fire Department, California

Captain Peterman and his crew returned from fighting a wildland fire at approximately 1900hrs. At approximately 2130hrs, firefighters noticed that Captain Peterman was sick. He was transported by fire department ambulance to the hospital and diagnosed as having had a heart attack.

Captain Peterman underwent emergency bypass surgery. He suffered multiple-system failure after the surgery and never recovered. Captain Peterman died on January 17, 2004.

APPENDIX B

FIREFIGHTER FATALITY INCLUSION CRITERIA – NATIONAL FIRE SERVICE ORGANIZATIONS

The National Fire Protection Association (NFPA), the National Fallen Firefighters Foundation (NFFF), the United States Fire Administration (USFA), and other organizations collect information on firefighter fatalities in the United States. Each organization uses a slightly different set of inclusion criteria that are based at least in part on the purposes of the information collection for each organization and data consistency.

As a result of these differing inclusion criteria, statistics about firefighter fatalities may be provided by each organization that do not coincide with one another. This section will explain the inclusion criteria for each organization and provide information about these differences.

The USFA includes firefighters in this report who die while onduty, become ill while onduty and later die, and firefighters who die within 24-hours of an emergency response or training regardless of whether the firefighter complained of illness while onduty. The USFA counts firefighter deaths that occur in the 50 States, the District of Columbia, and United States protectorates such as Puerto Rico and Guam. Detailed inclusion criteria for this report appear starting on page 96 of this report.

For 2004, the USFA reported 117 onduty firefighter fatalities.

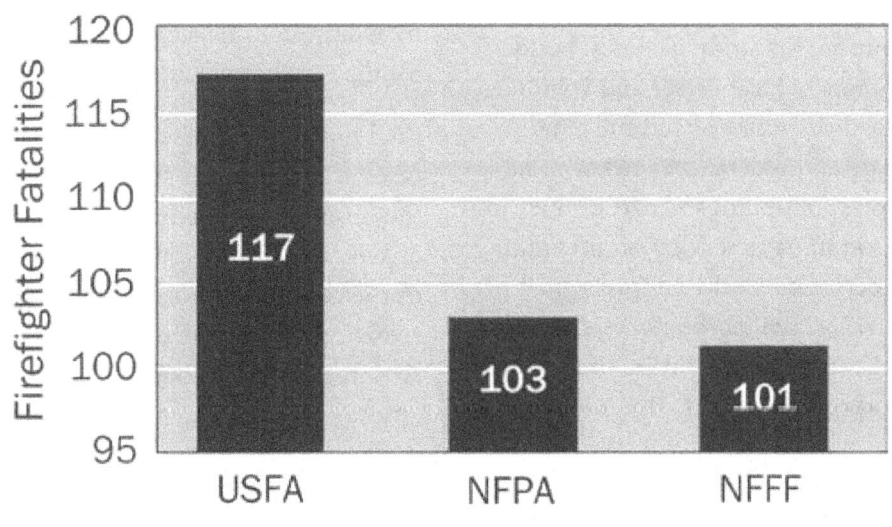

Firefighter Fatalities in 2004

INCLUSION CRITERIA FOR NFPA'S ANNUAL FIREFIGHTER FATALITY STUDY

Introduction

Each year, the National Fire Protection Association (NFPA) collects data on all firefighter fatalities in the United States that resulted from injuries or illnesses that occurred while the victims were onduty. The purpose of the study is to analyze trends in the types of illnesses and injuries resulting in death that occur while firefighters are on the job. This annual census of firefighter fatalities in its current format dates back to 1977. (Between 1974 and 1976, NFPA published a study of onduty firefighter fatalities that was not as comprehensive.)

What is a Firefighter?

For the purpose of the NFPA study, the term firefighter covers all uniformed members of organized fire departments, whether career, volunteer or combination, or contract; full-time public service officers acting as firefighters; State and Federal government fire service personnel; temporary fire suppression personnel operating under official auspices of one of the above; and privately employed firefighters including trained members of industrial or institutional fire brigades, whether full- or part-time.

Under this definition, the study includes, besides uniformed members of local career and volunteer fire departments, those seasonal and full-time employees of State and Federal agencies who have fire suppression responsibilities as part of their job description, prison inmates serving on firefighting crews, military personnel performing assigned fire suppression activities, civilian firefighters working at military installations, and members of industrial fire brigades. Impressed civilians also would be included if called on by the officer in charge of the incident to carry out specific duties. The NFPA study includes fatalities that occur in the 50 States and the District of Columbia.

What does 'onduty' mean?

The term onduty refers to being at the scene of an alarm, whether a fire or non-fire incident; being en route while responding to or returning from an alarm; performing other assigned duties such as training, maintenance, public education, inspection, investigations, court testimony, and fundraising; and being on call, under orders or on stand-by duty other than at home or at the individual's place of business. Fatalities that occur at a firefighter's home may be counted if the actions of the firefighter at the time of injury involved firefighting or rescue.

Onduty fatalities include any injury sustained in the line of duty that proves fatal, any illness that was incurred as a result of actions while onduty that proves fatal, and fatal mishaps involving nonemergency occupational hazards that occur while onduty. The types of injuries included in the first category are mainly those that occur at an incident scene, in training, or in accidents while responding to or returning from alarms. Illnesses (including heart attacks) are included when the exposure or onset of symptoms are tied to a specific incident of onduty activity. Those symptoms must have been in evidence while the victim was onduty for the fatality to be included in the study.

Fatal injuries and illnesses are included even in cases where death is considerably delayed. When the onset of the condition and the death occur in different years, the incident is counted in the year of the condition's onset. Medical documentation specifically tying the death to the specific injury is required for inclusion of these cases in the study.

Categories not included in the study

The NFPA study does not include members of fire department auxiliaries; nonuniformed employees of fire departments; emergency medical technicians who are not also firefighters; chaplains; or civilian dispatchers. The study also does not include suicides as onduty fatalities, even when the suicide occurs on fire department property.

The NFPA recognizes that a comprehensive study of firefighter onduty fatalities would include chronic illnesses (such as cardiovascular disease and certain cancers) that prove fatal and that arose from occupational factors. In practice, there is as yet no mechanism for identifying onduty fatalities that are due to illnesses that develop over long periods of time. This creates an incomplete picture when comparing occupational illnesses to other factors as causes of firefighter deaths. This is recognized as a gap the size of which cannot be identified at this time because of the limitations in tracking the exposure of firefighters to toxic environments and substances and the potential long-term effects of such exposures.

2004 Experience

In 2004, a total of 103 onduty firefighter deaths occurred in the United States, according to NFPA's inclusion criteria.

INCLUSION CRITERIA FOR NFFF'S ANNUAL FIREFIGHTER FATALITY STUDY

In 1997, fire service leaders formulated new criteria to determine eligibility for inclusion on the National Fallen Firefighter Memorial. Line-of-duty deaths shall be determined by the following standards:

1. (a) Deaths of firefighters meeting the Department of Justice's Public Safety Officers' Benefits (PSOB) program guidelines, and those cases that appear to meet these guidelines whether or not PSOB staff has adjudicated the specific case prior to the annual National Fallen Firefighters Memorial Service; and

   (b) Deaths of firefighters from injuries, heart attacks, or illnesses documented to show a direct link to a specific emergency incident or department-mandated training activity.

2. While PSOB guidelines cover only public safety officers, the Foundation's criteria also include contract firefighters and firefighters employed by a private company, such as those in an industrial brigade, provided that the deaths meet the standards listed above.

3. Some specific cases will be excluded from consideration, such as deaths attributable to suicide, alcohol or substance abuse, or other gross abuses as specified in the PSOB guidelines.

The National Fallen Firefighters Memorial was built in 1981 in Emmitsburg, Maryland. The names listed there begin with those firefighters who died in the line-of-duty that year. The United States Congress created the National Fallen Firefighters Foundation to lead a nationwide effort to remember America's fallen firefighters. Since 1992, the tax-exempt, nonprofit Foundation has developed and expanded programs to honor our fallen fire heroes and assist their families and coworkers by providing them with resources to rebuild their lives. Since 1997, the Foundation has managed the National Memorial Service held each October to honor the firefighters who died in the line-of-duty the previous year.

During October 2005 Memorial Weekend, the Foundation will be honoring 107 firefighters who died in the line of duty. The following is a list of those to be honored:

Gary Archibeque, Firefighter

Grady Roy Austin, Captain

Jeffrey C. Bergstrom, Firefighter/Paramedic

William E. Bierbower, Firefighter

Barbara L. Bordenkircher, Firefighter

Connie C. Bornman, Firefighter/EMT

Bob L. Boyles, Jr., Firefighter *

Charles G. Brace, Battalion Chief

Steven C. Brack, Firefighter

Herbert C. Caldwell III, Firefighter

Donald N. Carlson, Firefighter

Harold D. Chappell, Firefighter

Edward P. Conricote, Firefighter

Thomas Conway, Fire Police Captain

Brenda D. Cowan, Lieutenant

Patrick J. Cramer, Firefighter

Mario F. Cunha, Fire Engineer

Elliott Davis, Jr., Fire Commissioner

Robert J. DeAngelis, Firefighter *

Thomas DeAngelis, Captain

James D'heron, Deputy Fire Chief

Michael R. Dunlap, Firefighter

Terri L. Eiland, Lieutenant

Daniel E. Elkins, Captain

Steven W. Fierro, Firefighter

Keith A. Firment, Captain

Jaime L. Foster, Firefighter

Gary D. Fox, Fire Chief

Cordell W. French, Firefighter/EMS Director

Richard L. Gabrielli, Fire Police Officer

Glenn C. Galderisi, Firefighter

Leslie W. Gant, Jr., Lieutenant

Thomas Chester Gentry, Firefighter *

Jackson H. Gerhart, Firefighter

Leslie Keith Gillum, Firefighter

Robert A. Griffin, Firefighter

Irwin E. Gross, Firefighter

Willie Grudzinski, Assistant Chief

Nito Rene Guajardo, Firefighter

Derrick T. Harvey, Lieutenant

Charles T. Hatch, Jr., Firefighter/Paramedic

Ernest Heatherman, Fire Police Captain

Robert D. Heighton, Firefighter/Paramedic

Robert E. Heminger, Captain

Randy R. Henderson, Forestry Technician

Lawrence J. "Larry" Hoffman, Unit Protection Supervisor

Daniel P. Holmes, Forestry Tech - Hotshot

Jeffrey W. Howell, Firefighter

Phillip S. Hulen, Firefighter

Richard A. Jones, Firefighter

Michael J. Kilpatrick, Assistant Fire Chief

Kevin W. Kulow, Probationary Firefighter

Willie J. Lacy, Firefighter

Benjamin Matthew Lang, Firefighter/EMT

Charles M. Lehnen, Assistant Chief

William J. Lightbody, Rescue Member/Firefighter

Kenneth W. Lipyance, Lieutenant

Michael E. Lynch, Firefighter

David A. Mackie, Firefighter/Paramedic

Michael Stokes Martin, Firefighter

Michael McAdams, Firefighter

Gerald K. McGowan, Acting Captain

Kevin R. McIntyre, District Chief

Lewis R. McNally, Assistant Fire Chief

Mark E. Miller, Lieutenant

Jared Michael Moore, Firefighter

Theodore A. Myhre, Firefighter

Bret E. Neff, Deputy Chief

Barrie J. Niebergall, Driver/Operator/Engineer

Joseph S. Northup, Sr., Firefighter

Richard O'Brien, Firefighter

James Harold Pennington, Fire Chief

Joseph S. Pepe III, Firefighter

Raymond L. Peterman, Fire Captain

Edward O. Peters, Forest Ranger

Lester Phillips, Fire Chief

George H. Raber, Crew Chief

Edgar B. Rogers, Lieutenant

James L. Rogers, Fire Chief

Clinton L. Romine, Firefighter

Steven M. Rosenfeld, EMS Captain

Jason T. Rowe, Fire Chief

Rey Rubio, Firefighter

George O. Sandfield, Firefighter *

Paul Parsons Satterfield, Battalion Chief *

Eva M. Schicke, Firefighter

Kevin M. Shea, Fire Chief

Kevin L. Slain, Captain

Robert L. Smith, Firefighter

Edward "Ed" Stallings, Fire Chief

Richard A. Stefanakis, Master Firefighter

Kenneth E. Sterling, Firefighter/EMT

John J. Stoudt, Chief Engineer

Lawrence J. Sweetnich, Firefighter *

John D. Taylor, Jr., Captain

Kenneth M. Temke, Firefighter

Gary A. Tilton, Fire Chief

Alan D. Toepke, Firefighter

James E. Towell, Pilot

Wayne C. Turner, SEAT Pilot

David E. Vinisky, Firefighter

Donald Eugene Ward, Firefighter

Mike Ward, Pilot

Charles C. Webb, Lieutenant

William Weborg, Assistant Fire Chief

Rick L. Wilbur, Fire Chief

Kenneth A. Woitalewicz, Captain

*- Denotes a previous year fatality being honored this year.

APPENDIX C

HOMETOWN HEROES
SURVIVORS BENEFITS ACT OF 2003

PUBLIC LAW 108–182—DEC. 15, 2003   117 STAT. 2649

Public Law 108–182
108th Congress

An Act

To ensure that a public safety officer who suffers a fatal heart attack or stroke while on duty shall be presumed to have died in the line of duty for purposes of public safety officer survivor benefits.

Dec. 15, 2003

[S. 459]

Be it enacted by the Senate and House of Representatives of the United States of America in Congress assembled,

Hometown Heroes Survivors Benefits Act of 2003.
42 USC 3711 note.

SECTION 1. SHORT TITLE.

This Act may be cited as the "Hometown Heroes Survivors Benefits Act of 2003".

SEC. 2. FATAL HEART ATTACK OR STROKE ON DUTY PRESUMED TO BE DEATH IN LINE OF DUTY FOR PURPOSES OF PUBLIC SAFETY OFFICER SURVIVOR BENEFITS.

Section 1201 of the Omnibus Crime Control and Safe Streets Act of 1968 (42 U.S.C. 3796) is amended by adding at the end the following:

"(k) For purposes of this section, if a public safety officer dies as the direct and proximate result of a heart attack or stroke, that officer shall be presumed to have died as the direct and proximate result of a personal injury sustained in the line of duty, if—

"(1) that officer, while on duty—

"(A) engaged in a situation, and such engagement involved nonroutine stressful or strenuous physical law enforcement, fire suppression, rescue, hazardous material response, emergency medical services, prison security, disaster relief, or other emergency response activity; or

"(B) participated in a training exercise, and such participation involved nonroutine stressful or strenuous physical activity;

"(2) that officer died as a result of a heart attack or stroke suffered—

"(A) while engaging or participating as described under paragraph (1);

"(B) while still on that duty after so engaging or participating; or

"(C) not later than 24 hours after so engaging or participating; and

"(3) such presumption is not overcome by competent medical evidence to the contrary.

"(l) For purposes of subsection (k), 'nonroutine stressful or strenuous physical' excludes actions of a clerical, administrative, or nonmanual nature.".

Approved December 15, 2003.

LEGISLATIVE HISTORY—S. 459:
CONGRESSIONAL RECORD, Vol. 149 (2003):
    May 15, considerd and passed Senate.
    Nov. 21, considered and passed House, amended.
    Nov. 25, Senate concurred in House amendment.

○